D0026322

PRACTICAL ASTRONOMY:
A User-friendly Handbook
for Skywatchers

Then felt I like some watcher of the skies, when a new planet swims into his ken
John Keats

Dedicated to Ida my wife for her devoted help and forbearance throughout.

PRACTICAL ASTRONOMY:
A User-friendly Handbook
for Skywatchers

H. ROBERT MILLS, OBE, MSc, MIEE, CEng, DipEd, FRAS

Formerly Director of Science and Engineering, The British Council, London

Albion Publishing
Chichester

First published in 1994 by
ALBION PUBLISHING LIMITED
International Publishers, Coll House, Westergate Street, Aldingbourne, Chichester,
West Sussex, England
USA: **PAUL & COMPANY PUBLISHERS' CONSORTIUM INC.**, P.O Box 422,
Concord, MA 01742

British Library Cataloguing in Publication Data

Mills, H . Robert
Practical Astronomy: User-friendly
Handbook for Skywatchers
I. Title
522

ISBN 1-898563-02-0(Albion Publishing) Library Edition

ISBN 1-898563-00-4 (Albion Publishing) Paperback Edition

Typeset by Warren White Typesetting, Chichester
Printed in Great Britain by Hartnolls, Bodmin, Cornwall

Table of Contents

FOREWORD by Heather Couper 5

PREFACE 7

ACKNOWLEDGEMENTS 11

1. THE CELESTIAL SPHERE

1.1   Adapting to the dark 13

1.2   City lights 13

1.3   How may stars can we see? 14

1.4   The Pole star 16

1.5   Latitude 16

1.6   The Earth's axis 17

1.7   Star globes and maps 18

1.8   Star positions using a globe 19

1.9   Measuring angles 25

1.10   A quadrant for measuring altitudes and azimuths 29

1.11   The marine sextant 37

1.12   Sidereal time 39

1.13   Telling the time by the stars 44

1.14   A little Mathematical fun with Venus 50

1.15   A classroom model 52

1.16   Diagram of the celestial sphere 53

1.17   Finding the radius of the earth 55

1.18   A DIY model of the celestial sphere 57

1.19   A 3D starfinder and identifier 60

2. THE SUN AND SUNDIALS

2.1   The Sun as a timekeeper 63

2.2   The Copernicus revolution in astronomy 65

2.3   Sun spots 66

2.4 Sundials 71

2.5 Sundial designs 76

2.6 The altitude sundial — How the Sun's altitutde can tell us the time. 78

2.7 A calculation from the 14th Century 82

2.8 Making a vertical stick sundial 83

2.9 The Azimuth dial 86

2.10 Sundials and the equation of time 94

2.11 To measure the eccentricity of the Earth's orbit round the Sun . . . 96

2.12 Noon marks 99

2.13 The blue of the sky 100

2.14 Getting a bearing or checking a compass from the Sun 101

2.15 Unusual variety of an equatorial sundial 104

2.16 A portable polar sundial 106

2.17 Eclipses from the astronomy of Stonehenge 110

3. STAR POSITIONS, STAR MAPS, PLANISPHERES AND NOMOGRAMS

3.1 The need for spherical trigonometry 114

3.2 Finding and identifying stars 114

3.3 Computers and calculators 116

3.4 A note on planispheres and how they can be put to maximum practical use using an Alt-Az Graticule. 121

3.5 Type II planispheres stereographic projection 125

3.6 Stars of the month and their special maps 131

3.7 Using graphs and nomograms to solve problems of star positions for sky watchers, and for navigators 134

3.8 Star paths across the sky. 137

4. LIGHT AND BASIC OPTICS

4.1 Reflection 139

4.2 Refraction of light 140

4.3 Telescopes 144

4.4 The equatorial telescope 150

4.5 Refractors and reflectors 150

4.6 Using telescopes 154

4.7 Setting circles 155

4.8 The Moon 157

4.9 Lens formula for eyepiece projection 158

4.10 Taking photographs with your telescope 160

4.11 Magnification by eyepiece projection 160

4.12 Field of view in an eyepiece 162

4.13   Binoculars.................................... 163
4.14   The Barlow lens............................. 164
4.15   Observing with ease and comfort using a pair of binoculars
        or a small telescope.................................... 167
4.16   Rainbows 171
4.17   Light intrusion the bane of skywatchers 173
4.18   Spectra of stars 174
4.19   What a telescope can do under ideal conditions and
        related formulae................................... 176

5.   MISCELLANEOUS CALCULATIONS
5.1   The universal law of gravity 178
5.2   Gravity and ships sailing at sea........................ 179
5.3   The pendulum 180
5.4   The Foucault pendulum........................... 181
5.5   Meteors and shooting stars............................ 183
5.6   The precession of the Earth's axis...................... 194
5.7   The Earth's axis and the plane of the ecliptic 194
5.8   Spherical trigonometry used in astronomy 195
5.9   To find the limiting magnitude of a telescope 199
5.10   The dip of the horizon at sea and the radius of the Earth 201
5.11   A deep breath................................. 202
5.12   Spherical and parabolic mirrors 203
5.13   An analysis of the equations of time..................... 210

6.   APPENDICES AND USEFUL INFORMATION
6.1   Advice for beginners............................. 215
6.2   Giving a talk 215
6.3   Glossary 218
6.4   A note on nomograms............................. 221
6.5   The right ascensions and declinations of some of the brightest
        stars.. 228
6.6   Useful information............................... 230
6.7   Suggested further reading and sources of information.......... 232
6.8   Tailpiece.................................... 233

INDEX 234

Foreword

by **Dr Heather Couper,** Professor of Astronomy, Gresham College, London

I don't think I've ever seen Robert Mills without a device. Let me explain: at astronomical gatherings, Robert is there in a corner (surrounded by people, of course), busily demonstrating some (apparently) fiendish complex — er, device. It could be a planisphere, a flat representation of the sky; it might be a navigational instrument like a sextant; a pendulum; even a modern version of the old Arab star-finder, the astrolabe. One day he turned up with a can of lager — only to turn it into a very ingenious sundial!

Robert Mills thinks astronomy is fun. And as a highly experienced educator, he also knows that it is arguably the most important of all the sciences. Most people, including the young, have an instinctive fascination for astronomy. Many can recite the order of the planets in the Solar System long before they're seven or eight. It's well-known that astronomy is a sugar-coated pill that lures youngsters into studying less immediately attractive sciences like physics and chemistry. And I needn't labour the point about how much we need young scientists in the future.

Astronomy now forms part of the Curriculum in most English and Welsh schools. All very well, if it helps produce our next generation of scientists, but how do you go about teaching it? If you're a teacher, you need all the help you can get — and this book will certainly give you some ideas. The great thing about this *Handbook* is that it's all hands-on. Some of the gadgets may *look* complicated, but persevere — because you'll find they're a marvellous way of making astronomy applicable and down-to-earth. But, I hear you say, "astronomers do it at night" — while schools "do it" during the day. However, dozens of Robert Mills' projects are exclusively designed for daytime work, and therefore ideal for the classroom.

For those already bitten by the astronomy bug, read on — there's plenty in here for you too. Quite apart from some very practical advice on telescopes, making measurements, using starcharts and understanding lenses, there's enough to keep you busy for a whole string of cloudy nights.

And that brings me back to a point I remember Robert making. ''What would life be like,'' he asked, ''if the sky were *perpetually* clouded, so that we saw no Sun, no Moon and no stars — in other words, no astronomy?'' And he went on to answer the question: ''We would then have no means of telling the time, no clocks, no calendars, no compass directions, no exploration, no navigation, no communal life''.

And who said that astronomy wasn't a practical subject?

Heather Couper
July 1993

Preface

"I listen and forget, I see and I learn, I do and I understand."

(Chinese proverb)

The object of this book is to help people of all ages and their teachers, to engage in skywatching as an enjoyable and worthwhile pursuit that can enliven and enrich programmes of other sciences including mathematics! Modern astronomy is involved to some degree in practically all branches of science and these can be listed under the three headings shown in the Table on page 3. This list is not complete and there are other systems of classification, but it supports the view that astronomy has a rightful and valuable educational role to play in modern curriculum development.

So, astronomy draws richly on the sciences in promoting our understanding of the Universe. Inevitably it reciprocates and feeds back valuable and stimulating contributions to the advancement of science, particularly in Physics, Chemistry, Mathematics, Geology and Meteorology. Our nearest star, the Sun, and stars in their various stages of evolution have provided us with exciting evidence for modern theories of atomic structure. The sun is a large nuclear pressure cooker that produces vast amounts of radiant energy and streams of atomic and subatomic particles. We learn through curiosity and interest by asking questions, seeking answers and, above all, by **doing.** Rudyard Kipling comes to mind here:

I keep six honest serving men
(They taught me all I knew);
Their names are what and why and when
and how and where and who.

Whilst reading and using this book we should not lose sight of the awe and wonder and mystery of our Universe. Astronomy is a subject fraught with many facts and theories that, to the human mind, are quite beyond our understanding or belief. The Big Bang theory, derived apparently from a mysterious extrapolation of specially devised Mathematical Physics, expects us to believe at the present time that all the matter in the Universe at some primaeval instant about eighteen thousand million years ago was concentrated at infinite density in a space of zero volume. This object, or singularity, of mass 10^{50} tonnes exploded and may reach the boundaries of space, as we are led to believe, some 5×10^{12} years later at speeds approaching that of light. Eddington, the great Astronomer, was once asked after a lecture on the vastness of our Universe, *"Sir, is it not a fact that astronomically speaking man is but an insignificant speck in the universe?"* To which Eddington replied, *"Astronomically speaking man is the Astronomer"*

Brother Lawrence, a Franciscan monk, described a religious person as one who

"Practices the presence of God".

A dedicated astronomer is correspondingly one who

"Practices the presence of the Universe"

So let us not be discouraged or over-awed but stimulated and heartened when we contemplate things that are at present beyond our understanding. When we attempt to do things as sky watchers we are participating in the oldest of the sciences.

A study of the history of mankind reveals how astronomy has profoundly influenced the course of civilisation from Neolithic times (about 100,000 years ago) when man began to use the stars to fix the times for people to sow and plant crops and for hunters and animals to migrate. For a moment, reflect what life would be like now if our sky had been permanently overcast by clouds obscuring completely the sun, moon and stars; we should have had no means of making calendars, fixing dates, holding festivals or telling the time. There would have been no widespread sense of direction, no exploration or navigation. One of the most exciting sea stories from the 1st century AD is that told by St Paul, about his voyage in a grain ship from Sidon to Rome, a voyage which nearly ended in complete disaster, because the navigators could not use the sun or the stars to get their bearings, as verse 20 of the 27th Chapter of Acts of the Apostles relates:-

".and when neither sun nor stars shone upon us for many days, all hope that we should be saved was taken away."

They could not use the sun or stars for navigating their ship. They were lost in the Mediterranean.

THE MAIN BRANCHES OF ASTRONOMY

Descriptive	Gravitational	Physical
History	Dynamics	Composition of stars
The Ancient world	Kepler's Laws	Abundance of elements
Megalithic sites	Newton's Laws	Origin of stars and planets
Calendars		Organic chemical molecules in space
Astrolabes	Orbits under central forces	The origin of life
Star catalogues	Prediction of positions	Evolution
Positions of all celestial bodies	Distances	Earth science
	Momenta	Geology
Instruments	Masses of celestial bodies	Age and stellar evolution
Telescopes:- optical, radio	Sizes	White Dwarfs, Red Giants
Cameras and photography	Densities	Heat and energy production
Sundials	Cosmology	Temperature of stars
Clocks-sidereal time	Relativity	Radioactivity
Navigation by stars	Gravitational forces in stars' interiors	Interstellar space
Astrolabes	Tides	Plasma-physics
Scientific electronic calculation		Nuclear chemistry
Computers		Fundamental particles
Stellar magnitudes		Spectroscopy The radiation
Colour		spectrum from g-rays to radio waves
Star clusters		Quasars
Galaxies		Pulsars – red shift
Earth science		Black holes
Geography		Time measurement
Plaets		Propagation of light
Astronomical objects in works of art		Doppler effect
The impact of astronomy on religious thought and philosophy		Relativity
		Aberration
The Big Bang or the Creator.		Magnetic and electric fields in space
	The solar wind	

These three branches are closely linked so that each one depends for its advance on research and discoveries in the other two.

In latitudes about 53° N (North) during midsummer, we have only about four hours of real darkness around midnight, and in mid-winter, although we have eighteen hours of dark sky watching, many opportunities for good ob-serving are ruined by adverse climatic conditions or bright city lights. *"Practical Astronomy"* attempts to provide a variety of things to make and study in those intervals when the sky is impenetrable or fingers frozen, so as to be ready for a full and rewarding outdoor practical programme with telescopes, binoculars, theodolites and star charts, as soon as conditions are favourable.

However astronomy can also be studied in a practical way during the daylight hours. the sun, our nearest star, can give us many things to do in *daylight* both in the classroom and at home that are instructive and easy to carry out, and provide a valuable insight into our place on earth — as a space ship concerning our place in the solar system, and the nature and positions of other stars in our Galaxy. Most young skywatchers look with envy on astronauts who are privileged to take trips in a space ship; but think for a moment. *We* are *now* on a spaceship! The Earth on which we are fortunate to be is a marvellous spaceship, in orbit round the Sun, with all 'mod cons' laid on for our comfort; oxygen, comfortable temperature and air pressure and gravity that suits life perfectly! We can often gaze with wonder and enjoyment all round at the stars, galaxies, sun, moon and planets. Enjoy the privilege of being on a spaceship.The activities described in this book, will include:

- Using the sun to mark our base line, the N–S meridian;
- The construction of sundials of various kinds;
- The times and positions of the Sun at its rising and setting;
- The tilt of the earth's axis;
- The Sun's irregularities in keeping good time;
- Sunspots, flares and the sun's spectrum;
- Finding and identifying stars.

Although skywatchers are often called 'star-gazers', and the sun is a star, it is for us a very special star — at the centre of our solarsystem (2.2) it is so near that **we must never gaze directly at it   for to do so would permanently damage our eyes.** The human eye contains a powerful and beautiful little lens that focuses the sun's light and heat on the retina. The heat unrestrained is sufficient to burn the delicate network of sensitive nerves. Try this powerful heating effect by using a magnifying glass to focus an image of the sun on a piece of paper: the paper will soon smoulder and burn. The sun's rays are so strong that if they were to fall on a lens-like piece of glass in a dry region, they could start a fire as has often happened.

No mathematics teacher need be short of examples or projects in astronomy for exercises with a calculator. Astronomy can, and should, be used as a basis for an integrated science programme in view of the present absorbing interest in space probes, moon landings, lasers, black holes, radio and television programmes on astronomy.

Astronomy, however, can be successfully introduced into a science programme only if presented and received with understanding and enjoyment. Beware of the disaster that follows bad teaching and learning without understanding as exemplified by the boy who in answer to the question:

"What causes the tides?"

wrote, without proper understanding:

"The tides are caused by the rays of the moon striking the surface of the sea at an angle of 23.5° Fahrenheit"!

Many of the *"Things to do"* in this section have explanatory notes that may appear to stray into subject areas that are only marginally the province of astronomy. This is in keeping with the expressed object of the book, and in the hope that teachers of other disciplines, listed under the heading *"Descriptive Astronomy"* in the Table on page 9, will draw extensively on astronomy for examples and projects which can be a source of interest and excitement to young people in this space age.

ACKNOWLEDGEMENTS

I would like to express my thanks to many members of the British Astronomical Association, and the Association for Astronomy Education who have given the encouragement and helpful advice on my assembly of topics for this Handbook, many of which have been demonstrated at meetings of these Associations. I am particularly indebted to Commander Henry Hatfield, past President of the B.A.A. and to Eric Zucker, Editor of the A.A.E. Journal and Newsletter. I am greatful to Eddy Butt for his help with photographs, to Myra Newton for her skillful typing, and to John Carden for his initial typesetting work.

I wish to thank Ellis Horwood of Albion Books for this friendly help and cooperation in getting this book published. This is the outcome of over sixty years of friendship from our early days in India (Madras) when we collaborated in producing several books on Science for schools in India: he with Macmillan's Madras House, India, and I with the University of Madras. After the War, he inspired and helped me to produce my book on Teaching and Training. Throughout the years we and our families have remained close friends.

1

The Celestial Sphere

1.1 ADAPTING TO THE DARK

It is a mistake on a clear evening to decide to dash outdoors suddenly from a well-lit living room to have a quick look at the stars. If we do so we shall be dis- appointed at first to see nothing except two or three of the very brightest stars. We may then decide that skywatching is an overrated pastime.

It is only after about ten minutes in the dark that our eyes grow fully accustomed to the darkness, and the pupils of our eyes will open from about 2 mm diameter to their full aperture of about 7 mm. Eyes then admit to the retina about twelve times more light than they could with 2 mm aperture. The area of the wide open pupil $\pi \times (3.5)^2$ is 12.25 times larger than the area $\pi \times (1)^2$ of the partly closed pupil.

If you study a cat's eyes, you will notice how its pupils are wide open in darkness and are reduced to a small slit in a bright light (see Figs 1.1.1 and 1.1.2). The aperture of your camera lens varies its light gathering power in a similar way. When you next go out from a brightly lit room to look at the stars, start your stop watch and note the time it takes before you can see clearly the hundreds of stars spread all over the sky. Some people's eyes take longer to become fully dark adapted than others.

During short intervals that are necessarily spent looking at a star globe or chart, or making notes, use a torch with a red glass cover, as a dark-adapted eye responds less to red light than to other colours of the spectrum.

1.2 HOW SKY WATCHING IS SPOILED BY CITY LIGHTS

By far the greatest hindrance that many sky watchers have to contend with is the light intrusion of the sky near towns by street or motorway lighting. In order to look at objects of special interest, it is worthwhile to get away from it all into the country by car, taking binoculars or portable telescopes than can be used in comfort sitting in a car! Fig. 4.17.1 suggests how this might be done. Astronomical Societies can help to alleviate this problem by supporting all efforts of local authorities to adopt and instal economical street lights that direct their light downwards instead of wastefully into the upper air and so spoiling our chances of looking at the wonders of the heavens.

Fig. 1.1.1. Cats eyes in daylight — the pupil is small.

Fig. 1.1.2. Eyes at night — the pupil is large.

1.3   HOW MANY STARS CAN WE SEE?

People, when asked to say how many stars can be seen with the unaided eye, often give a wide variety of answers such as millions, thousands, hundreds or ''No idea''. This suggests something both ''to do'' and ''to discuss''.

If you start counting stars haphazardly you will soon land in trouble because you have to spread your counting over the whole of the Celestial Hemisphere visible from your horizon to the zenith, and you will soon either lose your way, or miscount.

The task becomes much more manageable if you divide the celestial sphere into a number of small areas and take samples, using a very simple star counter such as the one shown in Fig. 1.3. A process known as ''sampling'' in statistics.

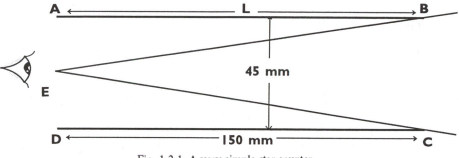

Fig. 1.3.1. A very simple star counter.

In Fig. 1.3, **ABCD** is a plastic tube such as a short piece of drainpipe 150 mm long of diameter 45 mm radius r.

The area of the sky covered in any one position is $\pi \times r^2$ as seen by the eye at **E**. The area of the whole celestial sphere $= 4 \pi L^2$ at Radius **L**. If, after a large number of counts representing all parts of the sky, it is found that the **average** number of stars that fall into each area of πr^2 is **n**, then assuming that the hemisphere that you cannot see has the same density of star distribution as the hemisphere that you have examined, a reasonable estimate of the total number of stars visible is

$$\frac{4 \pi L^2}{\pi r^2} \times n$$

In the Fig. **L** = 150 mm and **r** = 22.5 mm. Suppose **n** is 7. Then the number of stars visible is

$$= \frac{4 \times 150^2}{22.5^2} \times 7 = 1244$$

This exercise is not only something to do, but also to question and talk about:

- *Is this likely to be a reliable answer?*
- *What would make the result very inaccurate?*
- *What can be done to improve the estimated answer?*
- *Would several observers working on the project help?*

- *How to overcome different people's eyesight?*
- *How many samples would ensure a reasonable result?*

Careful counts show that the number of stars visible to the unaided eye is about 1500, but this number could vary between 1000 and 2000 depending on the observer's eyesight and on atmospheric conditions.

1.4 THE POLE STAR

We are now àware that there are about a thousand stars that can normally be seen, but that we are able to observe with **clear vision** only a very small patch of sky at a time. An important spot in the clear sky to look at for a start is due North and at an altitude above the horizon corresponding to your latitude. If your latitude is 51°N, then at 51° above the horizon you will see a rather lonely star which, although not very bright, has nevertheless played a significant part in world history, geography and navigation over the past several hundred years. It is known as **Polaris,** the Pole Star, it is a giant star at the great distance of 470 light years. We will start with this star as it never seems to move, and suggests one or two useful things to do.

1.5 LATITUDE

Our Earth is a planet which travels round the Sun once a year, roughly in an orbit of radius of 1.5 thousand million kilometers; furthermore it is also spinning on an axis which points within a fraction of a degree, to the Pole Star. This circumstance gives us not only the direction of North, but it enables us in navigation to find Earthly latitudes. We can best understand this by making use of an Earth Globe (acquired perhaps by amicable arrangement from the Geography Department) to simulate the spinning Earth by holding the model with its axis pointing to the Pole Star so that it is parallel to the Earth's axis as in Fig. 1.5.1. Now take a long knitting needle **AB** and position it with one end **A** touching your own geographical position, e.g., London, and the other end **B** pointing to the Pole Star. The needle is thus parallel to the axis of the Earth model. Now place a semicircular protractor with its centre point over London, keeping the straight edge **ED** in the plane **PAC**. Read the angle **EAB** made by the protractor's edge and the needle, **EAB** is the Latitude of London (51°N) and it is also the angle of the star above the horizon. **AD** points to the Southern horizon and **AE** points to the Northern horizon. Place the protractor in turn on Edinburgh (read its Latitude 56°N) and on Madrid to read its Latitude 40°N.

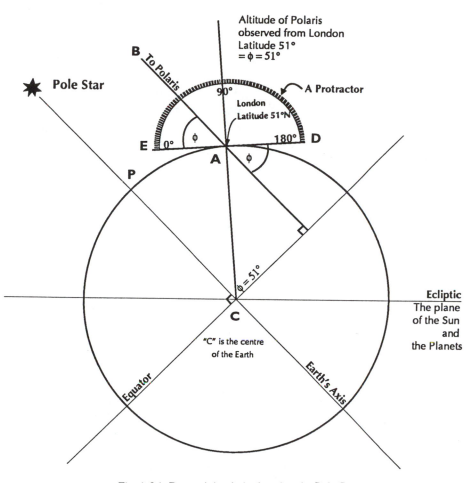

Fig. 1.5.1. Determining latitude using the Pole Star

1.6 THE EARTH'S AXIS

You may well ask *"How does the Earth's axis stay so steadily fixed in space, always pointing in the same direction?"*

The answer is that the massive Earth of about 6×10^{21} tonnes is spinning and that this spin keeps its axis steady like a top. You may point out that a spinning top does wobble a bit. True, and so does the Earth's axis; but its wobble is very slow and completes each full wobble round the pole of the ecliptic of about 47° across, in 25,800 years (see Fig. 5.6.1). So we, in a lifetime, do not live long enough to be aware of it. History, however, reminds us that the present Pole Star was not always in line with the Earth's axis and only in the last six or seven centuries has it been satisfactorily used to help navigators find their latitude.

William Shakespeare was wrong when he made Julius Caesar say boastfully:

"I am as constant as the Northern Star"

because at the time of Julius Caesar, there was no visible star in a constant position. Polaris, really Ursa Minor α which was part of the little Bear, was about 15° off the real celestial pole and so it went round the celestial pole in a circle 30° across, every 24 hours. Ursa Minor α has been moving nearer the celestial pole each century since then, and around 2,000 AD it will be at its nearest. Fig. 5.6.1 on page 194 shows that around 1,000 BC, Kochab — Ursa Minor β — was nearest to the celestial pole but it was of little use for navigation.

It must be said that Shakespeare was well versed in the astronomy of his day, 1564–1616, for then he recognised the Pole Star as a useful star for practical navigation. In his Sonnet, number 116, on *"True Love"* he compares true love with the seemingly fixed Pole Star and ends with these lines:

Love is not love
Which alters when it alteration finds
Or bends with the remover to remove.
Oh no! It is an ever-fixed mark
That looks on tempests and is never shaken.
*It is the star to every **wandering** barque*
Whose worth's unknown
*Although its **height be taken**.*

Taking the height (altitude) of a star in degrees above the horizon is the most important and useful measurement that can be taken of a star's position and it is there for the taking! The astrolabe was essentially an instrument for taking this height, and astrolabe means in Arabic "star taker" (see Fig. 3.5.6).

The sextant is an expensive, refined, precision instrument for taking star altitudes, so making various devices for taking altitudes and using them for star finding ranks high in our list of things to do. Instructions on how to make both an astrolabe and a sextant to measure altitudes of celestial bodies are given later in this Chapter.

Sky watchers soon become aware that all measurements in astronomy dealing with star positions are **angular measurements** made on the celestial sphere. Plane geometry has accordingly to give way to spherical trigonometry which in these days of the widespread use of calculators becomes delightfully simple.

1.7 STAR GLOBES AND MAPS

You may find it difficult to relate what you see on a star chart, or planisphere, with what you can see in the sky. Geographical maps represent to us shapes and positions of land objects to us as they would appear if we were actually looking down on them from a balloon or aircraft. Geographical maps are often drawn or checked this way from great heights by satellite.

Whenever we look at a star chart or planisphere, we find that we can relate the stars on the map to the stars in the sky only by holding the map above our heads between ourselves and the part of the sky we wish to look at. Most observers are able to perform this manoeuvre by using a little imagination, but the uncomfortable procedure gives the observer no precise information about the position of stars in terms of down to Earth measurements of altitude and bearing: that is the angles between the star and the horizon, and how far round the horizon the particular star is from the North point.

A correctly mounted **star globe**, set for latitude, date and time can give this information directly, as we shall see.

1.8   STAR POSITIONS USING A CELESTIAL GLOBE

In order to find our way among the stars in the sky, we have to understand how each star is given its own fixed position on the celestial sphere or on a star globe.

Fig. 1.8.1. A star globe constructed from a football.

Places on a geographical globe or map are fixed by using two coordinates: one is in degrees of latitude and the other is in degrees of longitude. Stars on a star globe are each allocated two analogous fixed coordinates that are called **degrees of declination** and **degrees of right ascension** respectively.

A star globe is a useful but rather neglected part of a skywatcher's equipment. It is expensive to buy, but instructive to use and particularly rewarding to make. They can be made from a football or a ballcock. The example shown in Fig. 1.8.1 is actually a plastic football, of the sort you can buy at any branch of Woolworths, which has been lightly sprayed with black emulsion paint. These footballs have a well defined 'equator' and two obvious 'poles' which serve a useful purpose when marking the declination circles that encircle the polar axis at intervals of 10° with declination at 0° at the equator and declination +90° at the North Pole and −90° at the South Pole.

The circle of the equator is divided into 24 parts each representing one hour of star time and marked 0^h, 1^h, 2^h, -24^h. This divides the globe into 24 hours of right ascension. As shown in Fig. 1.8.1, each division of 1 hour represents 15° of the globe's rotation. The zero mark for right ascension is the point on the equator where the Sun on the 21st March crosses the equator, denoted by the symbol ♈, which is at the point declination 0^h and Right Ascension 0^h.

In Fig. 1.8.2 is a ballcock globe showing its Declination and Right Ascension coordinates and is positioned in a bowl or large saucer, with its North pole pointing to the celestial pole.

Fig. 1.8.2. A star ballcock globe placed in a bowl.

1.8.2   The Star Globe

A correctly mounted star globe, set for a particular latitude and turned on its axis to the position required for the sidereal time on the date, can give a useful idea of the position of any star in terms of its altitude above the horizon as well as its bearing or azimuth. (sidereal time is discussed in Section 1.12.)

The star globe is mounted so as to turn about an axis through the globe from the South pole to the North pole (See Fig. 1.8.2). It is invariably positioned for practical use with this axis set parallel to the Earth's axis, i.e., pointing due north and tilted up from the horizontal by an angle equal to the latitude of the place. A star globe can be turned about its axis so that the position of the stars on the globe correspond to the positions of the stars in the sky as we should see them if we were in the centre of the globe, looking outward. This position depends of course on the date and the time, as the sky watcher is situated on a spinning and orbiting Earth.

1.8.3   A Sighting Tube

To assist sky watchers to derive full satisfaction from a 'set' star globe, take a piece of plastic tubing **T**, diameter about 35 mm, 300 mm long, with its ends cut carefully at right angles to the axis. Hold one end **A** firmly over a particular star on the globe, so that the tube lies in the direction of the radius of the sphere joining the centre to the star (see Fig. 1.8.3). Affix cross-wires of black cotton to the other end **B**. The tube thus points in the direction of the star, but does not enable us to look at it through the tube. This is easily overcome by drilling a small eye hole **H** about 10 mm diameter about halfway along the tube and cutting a 2 mm slot at an angle of 45 just below the eyehole and fitting a small rectangular piece of mirror **M** about 50 mm × 20 mm as in the Fig.. This enables the observer to find or identify any star by looking through **H** and seeing the star after reflection at the mirror.

The photograph in Fig. 1.8.4 shows the sighting tube **T** in position on the star globe which is set for Sidereal Time 14^h (marked on the globe). The position of the star by the azimuth circle is 75° and the altitude circle 40°. This position is that for the star Vega at Sidereal Time 14^h.

An easily made star globe can be made from a Woolworths' plastic football as shown in the photograph. The ball is suspended from its pole by a small bolt screwed carefully into the pump valve, and painted black with blackboard paint. When dry, 24 circles of right ascension and 12 circles of declination are drawn (analogous to longitude and latitude circles). These footballs have a well marked 'equator' which is a help when marking, as it is divided into 24 intervals each of 1 hour of Right Ascension.

In the photograph the globe is positioned in a bowl placed as shown on an Azimuth circular table.The axis of the ball is in the meridian and pointing to the pole star, parallel to the Earth's axis and inclined at an angle ϕ to the horizontal, where ϕ is the latitude.of the observer. Two uprights are screwed into the table on the North–South line of such a height that a piece of fine elastic **BC** stretched horizontally just touches the top of the globe at **Z** which is the zenith of the globe, and also helps the observer to identify the meridian and hence sidereal time. See Fig. 1.8.5.

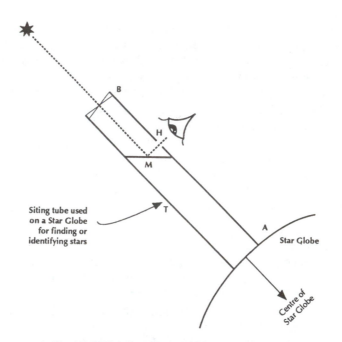

Fig. 1.8.3. Sighting tube in position on a star globe.

Fig. 1.8.4. A star globe with sighting tube set for Latitude 51° N, Sideral Time 14^h, the Star is Vega (Az 70°, Alt 40°).

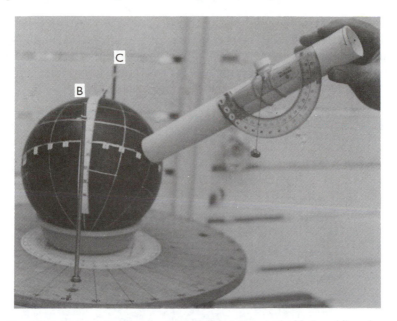

Fig. 1.8.5. A star globe whose uprights at B and C are in the N–S meridian plane.

It is important to note that an Earth globe is invariably mounted with its axis making an angle of 65.5° with the plane of the ecliptic, which we conveniently consider to be a horizontal plane.

The star globe axis however has to be tilted to suit the latitude of the observer, and as we have seen in section 1.5, it must make an angle φ with the horizontal where φ is the observer's latitude (e.g., 51° for London and 56° for Edinburgh).

1.8.6 A pocket Star Globe

We have considered some home-made globes, all using right ascension circles and declination parallels, on a particular sidereal time setting. These models can give us a star's position in familiar Earth terms, azimuth (bearing) and altitude above the horizon. Most of us appreciate having a pocket compass, so why not a pocket star globe? This challenge was fortuitously met by using a table tennis ball, A, and marking it with right ascension circles, and declination parallels, using a home made miniature lathe as shown in the Fig. 1.8.6.

Each ball has a well-defined meridian and a standard circumference of 120 mm which makes marking easy. The matt white surface takes erasable pencil markings nicely, as well as fine felt pen permanent ink markings. To make the markings the ball is held firmly between two pointed chucks which grip the ball at its poles, as shown in the Fig. 1.8.6. Fig. 1.8.7 shows the pocket globe positioned for Sidereal time 16^h in latitude 51° N, and the star **Altair**, RA 19h50m and declination 8°. The base stand B of the globe is a short piece of tube from a plastic pipe of 35 mm diameter which is marked round the base in degrees 0° to 360° to give bearings. Note the use of a sighting tube and protractor to give azimuths and altitudes. In Fig. 1.8.7, Altair is found to be bearing 110° and altitude 30° which agree approximately with the calculated values for this star's position at that time, and with results given by a planisphere and its graticule (see Chapter 3).

Fig. 1.8.6. A pocket star globe marked with right ascension circles and declination parallels by using a homemade lathe.

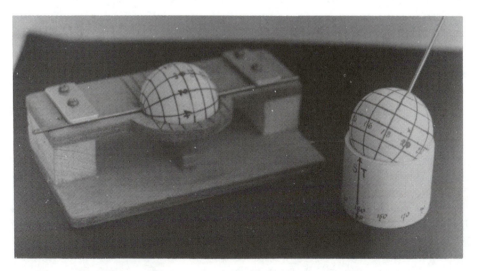

Fig. 1.8.7. A pocket star globe marked with Right Ascension circles and Declination parallels and mounted to show the Sidereal Time of 16^h in the meridian.

The table-tennis ball star globe makes no claim to be an accurate instrument, but it can be used conveniently to find where to look for a particular object, or to identify an object by revealing its approximate RA and Declination.

The D.I.Y. lathe can turn out four or five star globes in an hour. The miniature globes could make instructive gifts or items for sale at an astronomical society meeting.

1.9   MEASURING ANGLES

We have seen that the altitude of the Pole Star is the angle measured vertically, above the horizon, and that this angle is equal to the latitude of the observer.

In this section we have several simple ways for measuring angles between two objects on land or at sea. For very rough and ready measurements, hold your hand out at arm's length, then your fist makes an angle of about 10°. This can be greatly improved upon by holding upright a short ruler graduated in centimetres, and exactly at a distance of 57.3 cm from your eye. Then each centimetre on the scale makes an angle of 1° because a complete circle of radius 57.3 cm will have a circumference of $2\pi \times 57.3$ cm which is for our practical purpose = 360 cm which corresponds to 360° of the circle. To make the device more accurate and more convenient join the two ends of a plastic ruler **AB** by a piece of string 114.6 cm (2×57.3) long represented by **AEB** in Fig. 1.9.1. E is the mid-point of the string, (see Figs 1.9.2 and 1.9.3).

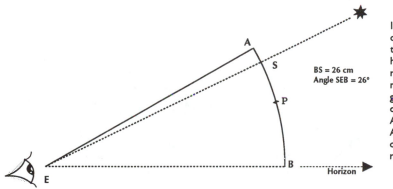

BS = 26 cm
Angle SEB = 26°

In use E is on the cheek just below the eye. EB is horizontal and the ruler is held at midpoint P and given a slight curvature so that AE = BE = PE. APB is then part of a circle of a radius of 57.3 cm.

Fig. 1.9.1. A simple device for measuring angles.

It is, of course, important for altitude measurements that the string **EB** should be horizontal. As we inhabit a world in which the vertical and horizontal are very familiar, **EB** can usually be estimated to be horizontal to within a degree or two, but for accurate measurements on land, when the horizon is not visible, and horizon point can be found to serve the purpose, by using a small spirit level as illustrated in Fig. 1.9.6.

Fig. 1.9.2. Measuring altitudes using a 30 cm ruler and string.

Fig. 1.9.3. Using the device for measuring azimuths.

For measuring angles greater than 30°, for example the altitude of the Pole Star as seen from London, note where the 30° mark comes in the sky, and then place the end **B** at that point and so carry on adding another 30°. The Pole Star will appear at the 21° mark — giving the altitude of the Pole Star as, 30° + 21° = 51°.

Similarly, bearings round the horizon from the North point can be measured in multiples of 30°. For example, 3 multiples from the North point will give you due East or due West, (90° from North).

A further refinement of these devices can be made by using a longer ruler, 50 cm long and by adding a wooden strut 57.3 cm long which can be clamped on a stand in the observing position, as shown in the Fig. 1.9.4 (this ensures that the arc **ADB** is circular). Measuring the altitudes of celestial objects is of great value in identifying them or finding them and in navigating by the stars, that is, in finding your own position on Earth.

A small spirit level fixed to the central strut at an angle of 25°, i.e., *parallel to the horizon string* **EB,** will ensure that altitudes are measured from the horizon. The observer's eye is at **E** and a star **S** will be seen at the point **H** which will register the angle **HEB**. At **E** is a small sighting hole (see Fig. 1.9.4).

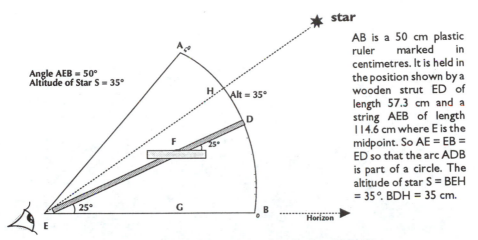

AB is a 50 cm plastic ruler marked in centimetres. It is held in the position shown by a wooden strut ED of length 57.3 cm and a string AEB of length 114.6 cm where E is the midpoint. So AE = EB = ED so that the arc ADB is part of a circle. The altitude of star S = BEH = 35°. BDH = 35 cm.

Angle AEB = 50°
Altitude of Star S = 35°

Fig. 1.9.4. A device for measuring altitudes from 0° to 50°.

This device has the advantage that it can be reversed so as to measure the altitude of the Sun **without looking at the Sun which must always be strictly avoided** (see Fig. 1.9.5). When used on the Sun, the string **BE** is pointed to the horizon with the point **A** below the string **BE** as in the figure. The Sun's altitude can then be recorded by the spot of light from the sighting tube falling on the scale, as shown in the photograph.

The 50 cm ruler thus comprises an angle **AEB** = 50°. The spirit level with its centre at **F** is fixed to the strut at an angle **DFH** of 25° and parallel to the string **AB** as shown, ensures that when the string **EB** is horizontal, then **EB** is pointing to the horizon. The device measures altitudes from 0° to 50°, but by accurately increasing the angle **DFH** from 25° to 50° the range of the altitudes that can be measured is increased to 75°.

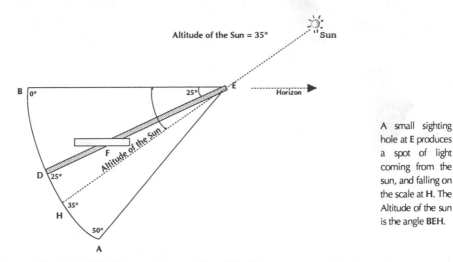

A small sighting hole at E produces a spot of light coming from the sun, and falling on the scale at H. The Altitude of the sun is the angle BEH.

Fig. 1.9.5. The ''Bow and Arrow Theodolite'' for measuring altitudes and Azimuths used as a Backstaff for observing the Sun.

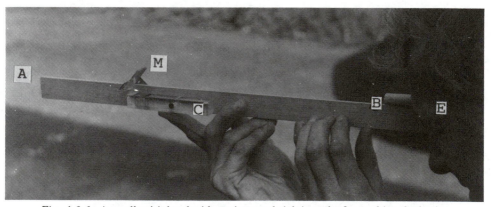

Fig. 1.9.6. A small spirit level with a mirror and sighting tube for marking the horizon. AB is a strip of wood with a spirit level at C, M is a small mirror that enables the eye of the observer at E to see the spirit level and ensure AB is horizontal. A small object directly in the line BA can be used as a point on the horizon for the purpose of measuring altitudes.

A 50 cm ruler is convenient as the arc **ADB** of 50° gives a quick way of spotting Polaris in latitudes in the around 50° latitude.

With the Sun in the direction of **HE** and the string **EB** pointing horizontally to the horizon, the spirit level remains in place parallel to **EB** Fig. 1.9.5.

When used for finding the altitudes of stars and particularly for the altitude of the Sun, the strut **DE** can conveniently be held in a laboratory stand.

The simple String and Ruler device Fig. 1.9.2 can be used to measure angles between stars, between Moon and planets. It can be used to measure horizontally along the horizon to find bearings, or azimuths, from the North point 0° or from the East, South or West points at 90°, 180° and 270° respectively as required.

An interesting group upon which to exercise the string and ruler 'theodolite' is the summer triangle formed by the stars Vega, Altair and Deneb.

Calculations using the formulae for spherical triangles show that the sum of the angles at Vega, Deneb and Altair = 187° which is greater than 180°. This need not cause surprise, as the sum of the angles of all spherical triangles is greater than 180°.

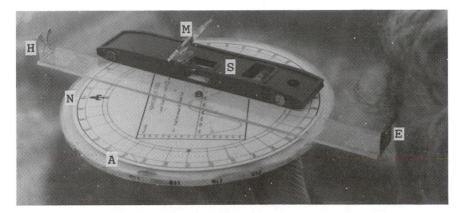

Fig. 1.9.7. An alternative device for locating the horizon, when making altitude measurements. EH is the horizontal line of sight towards the horizon, as decided by the spirit level S which can be seen from E in the 45 angled mirror M. Any suitable object on the line EH produced can be used as a point on the horizon.

We cannot measure the actual distance separating these stars but we can measure the angular distance between them on the celestial sphere. We can find out just where a star is on the celestial sphere if we know our own position on Earth, at a particular time of day and on a particular date. Conversely, if we know precisely where a star is in the heavens and if we know star time and the date, we can find where on Earth we are! This is the art and science of Astro-navigation which depends on knowing the Universal Time (UT) and being able to measure angles accurately. This will provide things to do in Chapter 3 using a calculator and graphical methods.

1.10   A QUADRANT FOR MEASURING THE ALTITUDE AND AZIMUTH OF ANY CELESTIAL BODY VISIBLE TO THE NAKED EYE

For those who may wish to exercise a little workshop skill, here is a a quadrant for measuring the altitude and azimuth of any celestial body visible to the naked eye.

In Fig. 1.10.1, **DOC** is a quadrant of a circle of radius 286.5 mm cut from a piece of good softboard 10 mm thick. **OD** and **OC** are two radii; the angle **DOC** is 90° and the curved length **DEC** is 286.5 × π/2 = 450 mm. A strip of centimetric graph paper 10 mm wide and 450 mm long is marked at 5 mm intervals so that each interval represents 1°. The strip is fixed to the quadrant with the 0° at **O** and the 90° mark at **D**. This forms the graduated altitude scale of the instrument. Each centimetre = 2°.

The quadrant is mounted on a wooden base 360 mm × 120 mm × 15 mm as shown in the figure. A line **AB** is marked along the centre of the base. **OM** is a plane mirror attached to the base. S is a spirit level that is used to ensure that the mirror and the base are horizontal when making an altitude observation. This is important. A strip of black plastic insulation tape is fixed across the centre of the mirror **OM** as shown in the figure. The edge marks the line on the mirror from which the reflected ray of light from the star emerges.

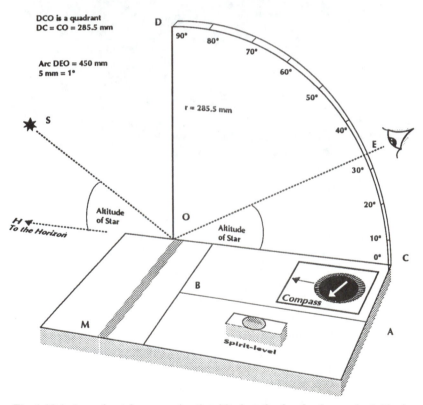

Fig. 1.10.1. A quadrant for measuring the altitude and azimuth of any celestial body visible to the naked eye.

To find the altitude of a star, point the line **AB** in the horizontal direction of the required star, and look for the reflected image of star in the mirror. The mirror is about 400 mm square and provides a convenient field of view for spotting or selecting the star. Move the eye so as to bring the reflected ray of light on to the edge **OM** of the black strip, and a few centimetres from **C**. Now use a fine needle or pin held horizontally across the graduated strip to locate exactly the point **E** which gives the altitude of the star **OCE** in degrees. The eye for comfortable observation should be at a distance for clear vision from **E** [250 mm].

To facilitate azimuth observations, an orienteering type of compass can be fixed to the base with its 'bearing' arrow parallel to the line **AB**. The compass can then read the bearing or azimuth of the star or body under observation, provided the appropriate correction has been made for the magnetic variation of the place, and no iron or steel nails or screws have been used in assembling the model.

Fig. 1.10.2. Transit posts marking the Meridian in the author's garden are useful for observing transits of the stars and planets at night or the Sun, Moon and times of Venus during the daytime

In Fig. 1.10.1, the N–S meridian line **AB** which marks the zero azimuth, is clearly important and a magnetic compass corrected for magnetic variation was used, but sky watchers for general observing will find it useful to mark in white emulsion paint a N–S line on a level plot of ground in the garden. Alternatively, a vertical pole about 2 m in length is set up at the Southern end of the plot and the shadow line cast by the pole is marked precisely at the time of the Sun's transit of the meridian, Fig. 1.10.2. This time is dependent on your longitude, and on the date as explained in Section 1.10.1. Thus the time is, 12 GMT –E (the equation of time + your Longitude West, or –Longitude East).

Fig. 1.10.3. The quadrant is here positioned on an azimuth circle and shows the altitude of the Sun to be 38° and its azimuth to be 122°. The nomogram pasted to the quadrant gives the Sun's Local Hour Angle as 42° and its declination to be 12°. The declination of the Sun is related to the date and has the value 12° on, or about, the 16th April. The Local Hour Angle of the Sun, 42°, represents a time of 42°/15 hours or 2h 48 m, before or after 12 noon. So the Sun time is 09h 12 m, or 14h 12m. The quadrant is thus a rough and ready clock and calendar, when used on the Sun.

Altitude and azimuth observations are what the ancient 'naked eye' astronomers were essentially concerned with, and what beginners today require when told where to look for any celestial object at a particular time. Instructive projects for school astronomy groups suggest themselves particularly if the instrument is used in conjunction with the set of altitude–azimuth curves, or nomograms, shown in Fig. 1.10.3 connecting altitudes, azimuths, declinations and hour angles.

Fig. 1.10.4 shows a means of taking altitudes using an alt-azimeter with a 45° mirror attached to a sighting tube that can be placed firmly on a table and angles from 0° to 90° can be measured with comfort. When viewing an object on the horizon, the sighting tube is vertical, when viewing an object at 90°, i.e., at the zenith, the tube is horizontal. This avoids getting pains in the neck or back. A plumbline from the protractor can measure the angle to within 0.5°. For use on the Sun a piece of paper **P** is placed at the end of the sighting tube **A** to receive the image of the Sun with the shadow of the cross-wires at the other end **B**. The device can be mounted on a wooden base which can turn about a pivot fixed at the centre of an azimuth circle, so that it can measure comfortably both the altitude and the azimuth of any object.

The sighting tube together with the mirror and the protractor are mounted on a piece of hardboard which can turn about a nut and bolt axis which passes through a vertical board that can turn freely about a horizontal base on an azimuth circle, **C**. The mirror **M** is fixed at 45° with respect to the sighting tube **ME**; **ME** is horizontal when observing objects at the zenith and is vertical when observing objects on the horizon. The sighting tube with mirror **M** and protractor are fixed to a piece of hardboard that can turn about an axis at **A**.

Fig. 1.10.4. An Alt-Azimeter with a mirror attached to a sighting tube at 45° enabling observation of stars overhead in comfort, as the sighting tube is then horizontal.

The altitude of a star above the horizon is a key measurement in astro-navigation which we can make in our own playground or backyard using devices described in this section. The altitude of stars for navigation and star-gazing is the natural and historical starting point for determining not only star positions, but also our own position on the Earth.

The story of trigonometry and spherical trigonometry goes back 2500 years to Pythagoras. His famous theorem about the square of the hypotenuse is the most important theorem of mathematics, particularly for astronomers, because it expresses a fundamental characteristic of space in which we live and relates the vertical and the horizontal, or the plumb line and the horizon (Fig. 1.10.4). Pythagoras gave numbers to geometry and astronomy which later led to trigonometry and, under the Arabs, to spherical trigonometry. Skywatchers eventually became fully aware that they lived on a sphere within a celestial sphere, and that star positions have to be measured and recorded in terms of angles marked on a sphere (see Section 5.8).

We have looked at several simple devices for measuring altitudes, correct to half a degree, but this is of limited value for navigators at sea: an error of half a degree represents an error of 30 nautical miles in the observer's position; not much comfort near a hazardous shore. It is nonetheless useful for finding or identifying stars, planets or comets. Altitudes and azimuths are co-ordinates to be made use of and enjoyed. Users of altitude-azimuth devices and instruments may be heartened to know that the large international telescope installed at La Palma, in the Canary Islands, is an altitude-azimuth instrument. With modern computerised controls it can hold stars in the field of view with the same precision, associated with equatorially mounted instruments.

1.10.5   Observing in comfort

Here is another simple yet versatile instrument that has the following uses and advantages for skywatchers: The swivel mirror slighting tube.

1. The device can be installed on the windowsill of a room or classroom, preferably at a window that faces in a southerly direction (see Section 2.11.1).
2. It is simple to construct and costs very little.
3. It does not employ a telescope or optical parts other than a small square of plane mirror, **M**, (Ladies' handbag pattern). This has inscribed upon it a thin black line **L** as shown in Fig. 1.10.5. **A** is a sighting tube consisting of a piece of plastic piping fitted with a 5 mm washer **E** as an eye piece at one end and with cross wires **F** at the other.
4. As illustrated, it can be used by an observer sitting down in comfort at a window which need not be open; this is a great asset in the Winter!
5. When aligned in the meridian, it can be used as an instructive transit instrument with which transit times can be recorded and translated into Sidereal Time (see Section 1.14). Section 1.10 The Sidereal Time of transit is equal to the RA of the star observed.

6. The mirror **M** is fastened to a small cylindrical axis **C** by a good adhesive and is free to turn in two bearings **D₁** and **D₂** consisting of a pair of small tool clips **D₁** and **D₂**. The mirror is able to turn and so deflect light from any celestial body along the sighting tube to the eye of the observer at **E**. The angle of the mirror with the horizontal can measure the altitude of the star observed in the cross wires **F** of the sighting tube by means of the pointer **P** coupling the mirror axis with a scale of angles calibrated from 0° to 90°. For example, when the mirror is horizontal it will reflect light from an altitude of 45° into the tube **EF** which is permanently fixed at an inclination of 45° on to a steel right-angled bracket as its support as shown in Fig. 1.11.5, using a good adhesive. In order to observe a star at the Zenith, i.e., at an altitude of 90°, the mirror is rotated on its axis anti-clockwise through an angle of 22.5°, and in order to observe a star on the horizon, the mirror is turned through an angle of 22.5° in a clockwise direction. The observer can thus scan the meridian from the horizon to the Zenith without moving the sighting tube or his or her position. The pointer, **P**, shows the star's altitude on the graduated scale.

Fig. 1.10.5. A simple versatile instrument, which can be used on a windowsill for altitudes, azimuths and transits.

7. It is portable and can be set up in any room or in the garden and if used in conjunction with an azimuth circle as shown, then the azimuths and altitudes of stars can be measured. For observations of the Sun, **which should never be looked at directly**, a small piece of tracing paper can be held at the end of the tube **E** and will indicate clearly as at **F** when the Sun is in the plane of the tube, and the pointer **P** is then recording the correct altitude of the Sun. Fig. 1.10.6 and 1.10.7.

    If the azimuth circle has been accurately set, then the line **L** on the mirror will be in the Sun's azimuth, as shown on the azimuth circle.

    The instrument can of course be used as a Sun Compass provided we observe the altitude of the Sun and know its declination, and then use the nomogram for the Sun shown in Fig. 3.7.3, which relates the four parameters Altitude, Declination, Hour Angle and Azimuth. It can also serve as a Sundial or Sun clock to find the Sun's local Hour Angle, or Local Sun Time, again by using the nomogram.

Fig. 1.10.6. Showing the device in use, observing the sun at the time of transit of the meridian.

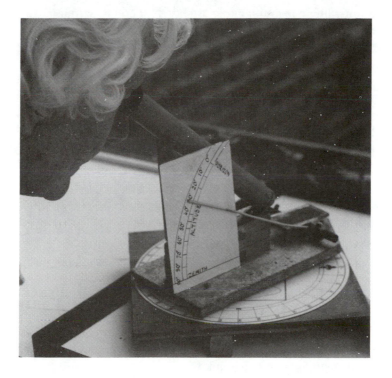

Fig. 1.10.7. The device in use on a windowsill facing due South.

1.11 THE MARINE SEXTANT

The Marine Sextant invented in the middle of the 18th Century made a great advance in navigation and was an instrument for measuring altitudes and other angles at sea, of optical and mechanical precision, accurate to 0.5 minutes of arc. It had the great advantage that it could be used at sea in rough weather because when a star's image is brought by mirrors to rest momentarily on the image of the horizon, the images of star and horizon remain in contact even with the ship heaving up and down.

Fig. 1.11.1 is a photograph of a simple D.I.Y. sextant. A photograph of a model made from ready to hand materials and costing very little, is shown with the various parts lettered as follows.

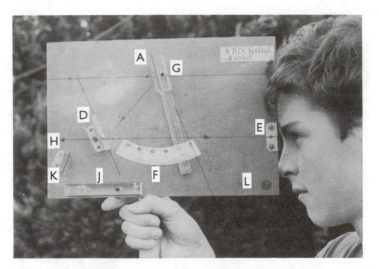

Fig. 1.11.1. A simple home-made sextant for demonstrations. it can be used for finding altitudes of objects or for angles between any two objects.

A   Index mirror of sextant.
BC   Index bar of wood or perspex, able to turn about a small bolt **G** as an axis. The bar is marked with a fine index line.
D   Horizon mirror about half the area of **A** and fixed in position.
E   Small hole 5mm diameter for sighting the horizon line **EH**.
F   A graduated arc of 60°. 1° of arc represents 2° of altitude or angle between two objects. This angle is independent of the ships movement.
J   Spirit level as a refinement to ensure that the line of sight **EH** is horizontal when the horizon is not visible.
K   Small inclined mirror 30 mm × 15 mm to facilitate view of the bubble of the spirit level from **E**.
L   Base plate of the model, made of hardboard 300 mm × 210 mm, with a handle **N** attached to the back.

The parts listed are mounted on the base board with a good adhesive or small nuts and bolts as appropriate after marking the layout in pencil.

To find your latitude from the altitude of Polaris you need to apply the corrections as navigators do.

As Polaris is not quite at the celestial pole, but 47′ away from it in the direction of the rather faint star Segin, ε Cassiopeiae (R.A. 2h), we can use this star to tell us the hour angle θ of Polaris at any time, and whether Polaris is above or below the celestial Pole, and how much. A protractor held steadily at arm's length will give you this angle to within a degree or two.

For example, an observation on Polaris on February 10th at 1700 hours local time will give a result of Latitude +47′. An observation made 12 hours later at 0500 hours [5 am, local time!] will give a result of Latitude −47′.

Then if we find the altitude H of Polaris with a sextant or any other accurate device, the correction to be made to this is to take off 47′ cos θ.

Then the correct height of the celestial pole, i.e., our latitude φ is H−47′ cos θ. We measure θ anti-clockwise from the vertical so where θ > 90°, cos θ becomes negative, so we have to add a little bit for correction. This is the principle used by professional navigators. the corrections for each hour and date of the year are given in tables but the single calculation above does the job very well.

You may ask ''is the correction necessary?''

Yes, if you are not to get lost or run aground because 47′ error in latitude means an error of 47 nautical miles, or 87 km, which could put your ship on the rocks!

An interesting exercise is to check this small circular wandering of Polaris round the celestial pole by measuring the altitude of Polaris when it is directly above the celestial pole, i.e., when the Sidereal Time is 2h 20m (1) and again when the Sidereal Time is 12 hours later, when the sidereal time is 14h 20m (2). Reference to a planisphere, or to the nomogram 1.13.4 (on page 47) which relates the date, the local mean time, and the Sidereal Time will give you convenient dates and times for making these observations.

1.12 SIDEREAL TIME

1.12.1 The value of a planetarium as a learning aid

If you get the chance to visit a planetarium, take it, as it is designed to show accurately the stars in their positions and movements and also (by special optical devices) the planets as they encircle the pole star. It puts you comfortably inside the star globe, and presents the celestial sphere as it appears on a fine night.

Planetaria are now established in many of the large cities of the world and provide not only special sessions in astronomy, as a part of general education, but also introductory courses in astro-navigation. The planetarium in its various forms has played an important role by presenting a working model of the celestial sphere which simulates with remarkable accuracy and clarity the star-filled sky and the apparent motions of all visible celestial bodies.

Planetaria from the days of simple globes and orreries of the 18th Century to the modern sophisticated optical devices have thus been effectively used in teaching astronomy and navigation. They help to clarify coordinate systems of stars on the celestial sphere and to promote a proper understanding of spherical triangles. Once these are fully understood the electronic calculator makes positional astronomy easy.

A small boy, city bred, was taken to a Planetarium and was duly thrilled by the spectacle of the night sky displayed on the celestial hemisphere. Later he attended his first camp in the heart of the country and saw the clear night sky unspoilt by street light for the first time and remarked with excitement, ''Cor! It's just like the Planetarium!''

If you can't get to a planetarium, here is something to do about it, as a simple simulation.

Take a large sized umbrella made from transparent plastic, open it and inscribe on the inside the more luminous stars and well known constellations in their correct Right Ascensions and Declinations given in almanacs and star books, using a ''permanent'' ink felt-tipped pen or small 'stick-on' stars that glow in the dark which are ideal for the purpose. The ribs of the umbrella will serve as useful Right Ascension lines and the plastic edge of the umbrella can be regarded as roughly representing the celestial equator (Fig. 1.12.1).

Fig. 1.12.1. A simulated mini planetarium, conisting of a transparent plastic umbrella on the inside of which are inscribed the more luminous stars and the main constellations in their correct Right Ascensions and Declinations. The stem of the umbrella is pointed to the celestial pole and the observer turns it anticlockwise to keep pace with the stars which appear to turn at a rate of 15° per hour. The umbrella device is severely limited as it gives approximate positions only for stars that have declinations greater than 50°. *See* the 3D starfinder which can be used in all latitudes and can include all stars. Section 1.19.

Now take the umbrella out on a dark clear night and provide yourself with a small hand torch with a red cover. After a few minutes for your eyes to become dark-adapted, point the shaft of the umbrella to the North celestial pole, and turn it about the shaft until stars marked on the umbrella coincide with those in the sky. Clamp the shaft to keep it steady, take a little walk, and after one hour you will have to turn the umbrella shaft through 15° anti-clockwise to catch up with the stars that appear to turn round the shaft. In fact the stars are not moving but the Earth is turning on its axis.

What you have been looking at, through the umbrella, is the working of a **Star Clock**, or **Sidereal Clock**, which runs very accurately at the rate at which the Earth spins on its axis.

If you have a tendency towards insomnia, you could keep watch under your transparent umbrella from 6 pm to 6 am the next morning during which time the stars would be seen to have moved 180° round the Pole. A far more comfortable and instructive way to keep watch is to let a camera do it for you. Set up an ordinary 35 mm camera on a secure stand, so that it points to the Pole Star. Use a time exposure of several hours, a wide aperture and a fast film such as Ektachrome 400. The result will show, in colour, the trails of stars as they move round the Pole, and the length of the arc of each star will be in angular measure, the exposure time in hours multiplied by 15° since the stars go round the Pole 15° each hour. Star trails can be made of stars in any part of the heavens, but the curvature of their arcs will depend on the declinations. Stars near the celestial equator [declination 0°] will have arcs that appear as straight lines on the film.

Fig. 1.12.2. A star globe, sighting tube and quartz clock with its crystal appropriate for Sideral Time.

Fig. 1.12.3. Star trials of circumpolar stars. The exposure time was 40 minutes.

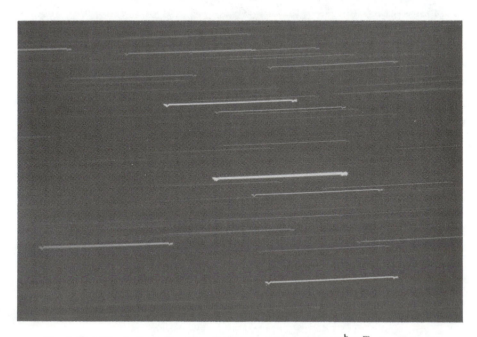

Fig. 1.12.4. Star trails in direction of Altair in α Aquilae RA $10^h 50^m$, also shown are
Aquilae β and γ. The exposure time was 13.7 minutes.

Photographing star trails can be a rewarding exercise, but it does require patience and experimental skill. Choose a dark clear night, moonless, and away from city lights, for best results.

The stars appear to make a complete turn of 360° round the pole star in about 24 hours, less four minutes.

The reason for this is that our Earth spins on its axis 366.24 times in one year when viewed *from the stars,* that is when we as sky watchers keep our eyes on the stars; but a person on the Sun will declare that the Earth has turned only 365.24 times — one turn, or one day less, than the star watcher's result. To demonstrate this loss of a complete turn, get a friend to stand in the middle of a room; then take a football with a white spot painted on it, and walk around your friend keeping the white spot always pointing in the direction of one and the same corner of the room. Your friend will say that the ball appears to have made a complete turn, but to anyone standing in the corner the ball has not turned at all.

This accounts for the fact that the Earth as seen by the stars turns 366.24 times in a year, but as seen by the Sun, turns only 365.24 times, and demonstrates that a star clock gains about 4 minutes each day so that in the course of a year of 365 days the star clock gains approximately,

$$\frac{365 \times 4}{60} \approx 24 \text{ hours.}$$

This rate of gain is easy to remember, as a sidereal clock gains about 10 seconds every hour. It is useful to note that the ratio of sidereal clock rate/mean Sun clock rate is

$$\frac{\text{Sideral Clock Rate}}{\text{Mean Sun Clock Rate}} = \frac{366.24}{365.24} = 1.0027379$$

So a sideral clock gains 0.0027379 of a day, each day or

$$0.0027379 \times 24 \times 60 \text{ minutes}$$
$$= 3.942576 \text{ minutes}$$

or 3 minutes 56.56 seconds per day.

The Greenwich Sidereal Time at midnight GMT can be calculated to within a minute or two by using the fact above that a Sidereal Clock gains 3.942576 minutes on our Mean Sun Clock each day that has elapsed since the 21st September when these two clocks record the same times approximately.

So on 10th December at midnight, 0^h, 80 days after 21st September, the Sidereal Time at Greenwich is

$$0^h + 80 \times 3.942576 \text{ minutes} = 5^h \, 15^m .$$

The Sidereal Clock for Sky Watchers can be reckoned to gain 10 seconds every hour, so Sidereal Time at 6am would be:

$$\text{Sidereal Time at } 0^h + 5^h \, 15^m + 6 \times 10 \text{ seconds} = 5^h \, 16^m .$$

an observer in longitude L [the Sidereal Time = $5^h 16^m + L^h/15$] would of course correct Greenwich for Local Sidereal Time by adding L/15 hours for Longitude L West or subtracting this if East.

To facilitate finding the number of days that have elapsed since a particular date, for the purpose of finding the approximate Local Sideral Time and then the local Hour Angle the following Table may be useful.

Table 1.12.1. Number of days since 21st September

To 1st October	10
To 1st November	40
To 1st December	71
To 1st January	102
To 1st February	133
To 1st March	161
To 1st April	192
To 1st May	222
To 1st June	253
To 1st July	283
To 1st August	314
To 1st September	345

1.13 TELLING THE TIME BY THE STARS

One of the things that young skywatchers are attracted to do is to tell the time by the stars. They perhaps look this topic up in a book on astronomy and become put off by a description of the complicated mediaeval nocturnal, and its difficult method of use and construction.

Fig. 1.13.1 shows a device that reads sideral time directly which can then be translated rapidly into local time by a simple graph, Fig. 1.13.4

The device makes use of the fact that a line between the stars β Cassiopeiae (Caph) and γ Ursae Majoris (Phecda) passes very nearly through the pole star, Polaris. This line, if imagined to have an arrowhead at Caph, makes an excellent ''hand'' for the sidereal clock face that surrounds the north celestial pole, with 0^h at the top.

The accuracy of this use of a Phecda–Caph line as a sidereal clock hand can be checked by using a polar map of the Northern sky, and joining Phecda and Caph by the straight edge of a ruler, as in Fig. 1.13.2.

The diagram, in Fig. 1.13.1, shows how the clock dial, hand, and sighting tube are assembled. Correct dimensions are given for those who may wish to make their own model. The sighting tube should be fixed perpendicular to the dial and fitted with cross wires. The clock hand is conveniently made of transparent plastic such as Plexiglas or Perspex, with a line scribed centrally along it. The hole in the middle of this strip should fit snugly on the sight, yet permit smooth rotation.

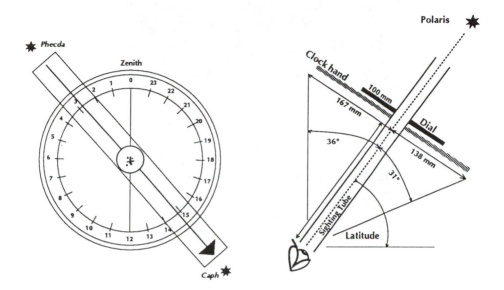

Fig. 1.13.1. Left: Reading Sideral Time Directly. Right: Assembly of Clock Dial, hand, and Sighting Tube.

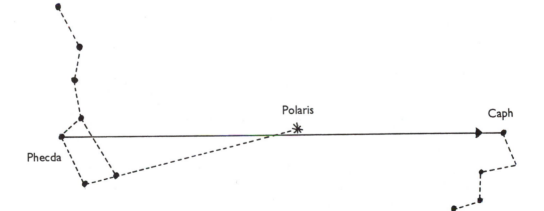

Fig. 1.13.2. Phecda-Caph Line as a Sideral Clock Hand.

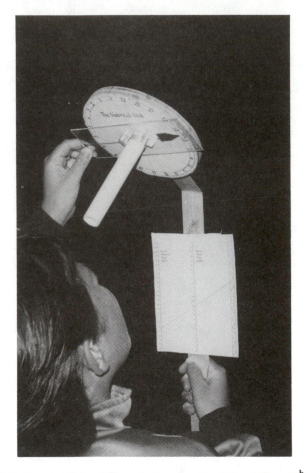

Fig. 1.13.3. Showing Sidereal Time device recording Sidereal Time 18^{h}30^m on 29th
May. This time is used on the nomogram in Fig. 1.13.4, which shows the local mean
time to be 02^{h}00 a.m.

To use this device to find Sidereal Time, observe Polaris through the tube and
align it with the cross wires. The 0^h– to –12^h line of the dial should lie in a vertical
plane, as determined by a plumb bob. The sighting tube of the device can be held
or clamped in position, parallel to the Earth's axis, and the dial lies in the equatorial
plane. When you turn the arrow to point at Caph, with the other end of Phecda,
local sidereal time can be read directly our the dial. The nomogram (or graph)
quickly converts sidereal to local mean time, as shown in Fig. 1.13.4 or this can be
done by using the date disc of Fig. 1.13.5.

The angular distances of Caph and Phecda from the pole are 31° and 36°, respectively. The eye can easily see these stars and Polaris at the same time without
appreciably changing position. To make and use this simple device can be an instructive and worthwhile project.

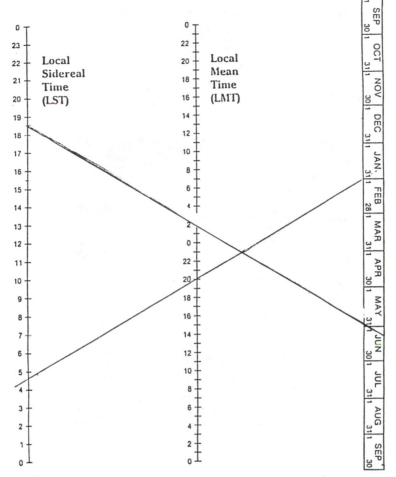

Fig. 1.13.4. A nomogram showing the local mean time on the 29th May.

We can use the nomogram in Fig. 1.13.4 to find the date and time at which any star, planet or celestial body of known Right Ascension will appear on the meridian with altitude [90 − φ + declination]. We can estimate the star's position one or two hours earlier or later to know where to look.

For simplicity, we can take the best viewing time as midnight because it is then dark all the year round between latitudes 65° N and 65° S, and stars can generally best be seen when they are near the meridian, their altitudes are at their maximum and their Right Ascension is equal to the Local Sidereal Time.

Example 1
What stars are in a favourable position for observing on 22nd February? In other words, what stars will be culminating at midnight on that date?

The nomogram shows this to be stars of Right Ascension near 10^h 0^m. This will help to identify the star using the star list in Fig. 5.15.1 as Regulus. The 21st's altitude will then be [90 − φ + 12°5′].

Example 2
When will a star of Right Ascension 6^h 44^m and declination −16° 41′ be in an ideal position for observation?

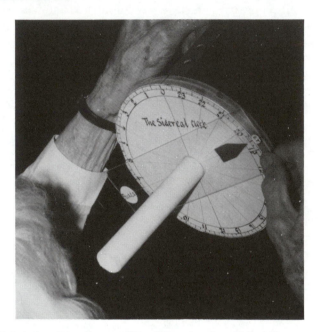

Fig. 1.13.5. A date disc conversts Sideral Time into Local Mean Time on any day of the year.

The alignment nomogram shows this star will be on the meridian at midnight on 31st December and its altitude is 90 − φ − 16° 41° i.e. 22° 19′ in latitutde 51°.

An instructive variation of the device for telling the time from the sidereal time of the stars is shown in Fig. 1.13.6. In this the nomogram has been replaced by a date disc which has the dates of the calendar marked round its periphery and is fixed to the Phecda–Caph pointer along the line joining the date 21st March and 21st September. It will be seen that when the hand is set on sidereal time, then the corresponding Local Mean Time 02^h is opposite the date of the observation. The Local Mean Time so found can be checked from the nomogram, or by a simple calculation using the fact that the Sidereal Time gains about 4 minutes (3.942676^m) for each day past the 22nd September. Using the example shown in Fig. 1.13.6 set for Sideral Time 18^h 30^m. The Nomogram for 29th May gives Local Mean Time as 02^h and the calculator gives the accumulated difference Sidereal

Time — Local Mean Time on 29th May, i.e., 250 days after 21st September as Sideral Time $-250 \times 3.942676/60$ hours $= 18^h\ 30^m - 16^h\ 30^m = 02^h$.

The photograph (Fig. 1.13.5) shows a sidereal clock face that is in the plane of the equator. The sighting tube **T** is aimed at the pole star. The hand of the star clock is **O**⊤ and is showing sidereal time $18^h\ 30^m$.

This is a useful device as the sidereal time on a particular date and at a known time tells you the Right Ascension of the star that is at that instant crossing your meridian. It also tells you what stars have just culminated and what stars to expect.

It is useful in applying the correction required when using the pole star to give an accurate latitude. From section 1.11. The correction is Latitude $\phi = H\text{-}47'\cos\theta$ where θ is the hour angle of Polaris = Sidereal Time − RA Polaris. Where the RA of Polaris is $2^h\ 23^m$ (see Fig. 1.16), and H is the observed altitude of Polaris.

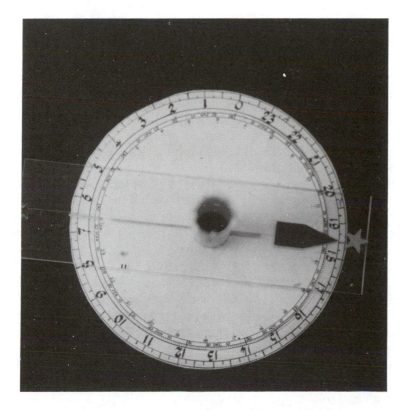

Fig. 1.13.6. Showing Sidereal Time 18^h30^m. The Local Mean Time indicated by the date, 29th May is 2 a.m. which confirms the time given by the nomogram of Fig. 1.13.4.

The clock shown in Fig 1.12.2 is a quartz clock which gains $3^m 56.55^s$ per day. You may wonder how it is possible to get a quartz clock (and most clocks are now quartz regulated) to gain 3 minutes 56.55 seconds per day. This provides something practical to do with a quartz crystal clock. The crystal vibrates at 4.19304×10^6 times per second or 4.19304 MHz to regulate our mean Sun clocks. To get the clock to gain 3 minutes 56.55 seconds per day the clock crystal has to vibrate at 4.205788 MHz to keep Sidereal Time.

$$\frac{\text{Sidereal Time Rate}}{\text{Mean Sun Time Rate}} = \frac{4.205788}{4.19304} = 1.002738 \tag{1}$$

Firms dealing in astronomical goods (see section 1.12) are well aware of the value of sidereal clocks and manufacture quartz crystals that vibrate at a frequency of 4.205788 mhz so all that is needed is to disconnect the normal crystal and substitute one of the higher frequency. The clock shown in Fig. 1.12.2 is one that has undergone this transplant operation that requires only a little skilled soldering.

1.14   A LITTLE MATHEMATICAL FUN WITH VENUS — TO FIND THE APPROXIMATE DISTANCE OF VENUS FROM THE SUN

Venus provides a special interest for skywatchers as it can appear as a brilliant evening star, and a few months later it appears as an equally brilliant morning star. It was not until Galileo with his telescope detected that Venus had phases similar to our Moon, that it became clear that Venus was going round the Sun, inside our Earth's orbit, and that it goes round the Sun in less time than the Earth does. Venus appears to engage in a space race against Earth round the Sun and has the great advantage of being in the "inside lane". After each lap by Venus round the Sun, Venus is ahead in angular distance, not only has it the advantage of the inner lane but it is also a faster mover.

Venus can do one lap, or orbit, in 244 days while Earth takes 366 days, Venus will therefore catch up with the Earth a lap ahead, or a gain of 2π radians after $225 \times 365/141$ days = 582 days. This is known as the **synodic period** of Venus.

Venus is especially interesting to sky watchers as it can be seen in broad daylight, especially when it is at its maximum elongation from the Sun when it has a 50% phase like a small half Moon.

Fig. 1.14.1 shows a way to measure the elongation angle **ESV** between Venus and the Sun. Here are some things to do about Venus using the angle measurer illustrated in Fig. 1.14.2.

Measure the maximum elongation of Venus as a morning star, then about 4.5 months later measure its elongation (following the Sun) in the evening; thus we achieve a figure as in Fig. 1.14.1.

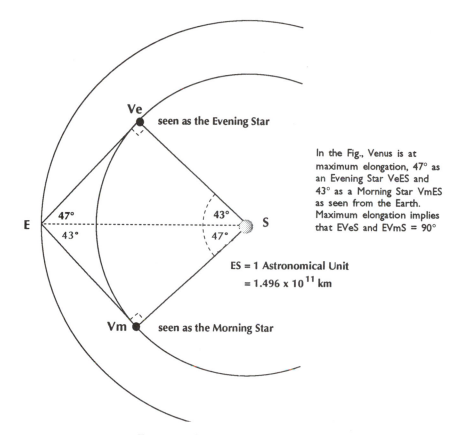

In the Fig., Venus is at maximum elongation, 47° as an Evening Star VeES and 43° as a Morning Star VmES as seen from the Earth. Maximum elongation implies that EVeS and EVmS = 90°

ES = 1 Astronomical Unit
= 1.496 x 10^{11} km

Fig. 1.14.1. Measuring the elongation of Venus.

Example

Suppose the maximum evening elongation $ESV_e = 43°$ and the maximum morning elongation $ESV_m = 47°$. Then if **ES** is the distance between Earth and Sun (1 Astronomical unit) = 1.496×10^{11} km. V_eS, the distance from Venus to the Sun = 1.496×10^{11} cos 43° = 1.094×10^{11} km (evening apparition).

For measuring the angle between two stars one requires only the use of a 300 mm plastic ruler and a length of string as shown in Fig. 1.9.2, but for measuring the angle between the Sun and Venus we must **on no account look into the Sun to make an observation**. This can be avoided by joining two strips of perspex each 300 mm × 25 mm as shown in Fig 1.14.2 **AB** has a fine small nut and bolt projection at **P**, while **CD** has a long projection at **Q**.

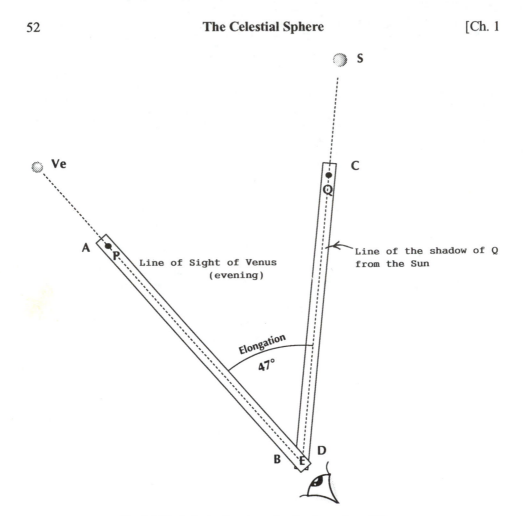

Fig. 1.14.2. A device for measuring the elongation of Venus.

When observing the eye is placed at E and is used to line up the arm **AB** on Venus, while **CD** is lined up on the Sun, not by eye, but by having the shadow of the projection **Q** fall along **QE** so that the angle **PEQ** is the elongation of Venus and the Sun. A shade should be arranged at **D** to ensure that light from the Sun does not fall directly onto the eye.

1.15 A CLASSROOM MODEL

A classroom model can be made using a football for the Earth and a tennis ball for the Moon. For actual dimensions see section 6.6.

Imagine the Earth reduced to the size of a football, which is about 20 cm in diameter (i.e., 0.2 m). The Earth has a diameter of 12.756×10^6 m. The **scale factor** for reduction of all distances is:

$$\frac{0.2}{12.756 \times 10^6} = 1.568 \times 10^{-8}$$

The Sun on this scale would have to be represented by a ball $2 \times 6.96 \times 10^8 \times 1.568 \times 10^{-8} = 21.8$ m in diameter, that is a little larger than a cricket pitch! and situated at a distance of the true distance of the Sun $\times 1.568 \times 10^{-8} = 1.5 \times 10^{11} \times 1.568 \times 10^{-8}$ m $= 2.35 \times 10^3$ m or 2.35 kilometres.

Apply the same scale to the Moon (radius 1.738×10^6 m).

The model Moon would be $2 \times 1.738 \times 10^6 \times 1.568 \times 10^{-8}$ diameter $= 5.45 \times 10^{-2}$ or 5.45 cm about the size of a tennis ball, which would be situated at a distance from the ''Earth'', given by:

$$\text{True distance} \times 1.568 \times 10^{-8} = 3.844 \times 10^{-8} \times 1.568 \times 10^{-8} = 6.03 \text{ m.}$$

To illustrate an **eclipse**, the tennis ball at 6.03 m will appear to be the same size (angular measure) as the Sun ball with the diameter of a cricket pitch 21.8 m. The model can also illustrate how and when eclipses occur, because if you look at a tennis ball at a distance of 6.03 m — in the direction of the model Sun, the tennis ball would just cover the 21.8 m sphere representing the Sun as in Fig. 1.15.1.

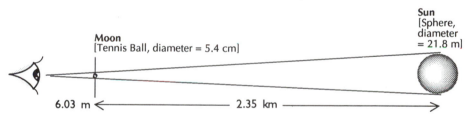

Fig. 1.15.1. An eclipse of the Sun by the Moon.

1.16 DIAGRAM OF THE CELESTIAL SPHERE, SHOWING THE TERMS USED

P   is the celestial pole — very near the Pole Star, Polaris.

Z   is the zenith — the point in the sky directly overhead.

Z′   is the nadir — the point in the sky which appears to be overhead to an observer situated at **Z′**.

X   is a star or celestial body being observed.

PZX   is a spherical triangle used to define star positions and important angles in navigation.

ZPX   is the hour angle of the star — it is the spherical angle at the pole **P** made by the star and your zenith. It is also measured by the distance **KQ** in degrees round the celestial equator.

XA   is the altitude of the star or angular distance of the star above the horizon.

A   is the point on the horizon where the circle from **Z** through the star, cuts the horizon. Azimuths are measured clockwise from N.

♈   is the point on the equator which has Right Ascension 0 or 0^h and is thus the point from which Right Ascensions are measured.

NA   is the angular distance which gives the bearing from North (measured N → E → S → W).

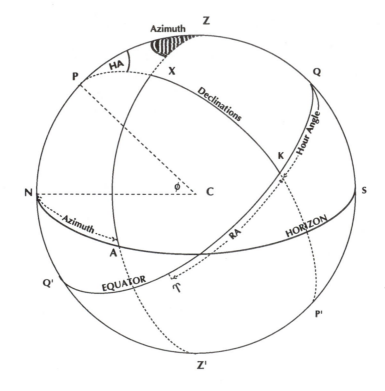

Fig. 1.16.1. A diagram of the Celestial Sphere.

When a star is culminating, i.e., it is in the meridian plane, then its hour angle is zero, and Local Sideral Time at that instant is equal to the Right Ascension of the star.

1.17 FINDING THE RADIUS OF THE EARTH

This can be done by making observations of the Sun's altitude in any country. This is something that two schools or groups (approximately in the same longitude) well separated by distance D km in latitude, can undertake with a little cooperation.

For example, consider a group in Southampton and another in Newcastle upon Tyne, 452 km apart. On a prearranged date during a spell of fine weather, each class measures the altitude of the Sun just as it crosses the local meridian. This can be done conveniently by marking the meridian line **N–S** on a level piece of ground, and by supporting a metre scale **AB** (100 cm from end to end) vertically over the line, using a plumbline. Measure accurately the length of the shadow **L** of the scale, cast by the Sun. The altitude of the Sun at transit in Southampton is A_S where tan A_S = **L** / 100. The altitude of the Sun at transit in Newcastle is A_N. The difference $A_S - A_N$ is the difference in latitude between Southampton and Newcastle and gives the distance between the two cities in degrees of latitude, and each degree of latitude represents 60 nautical miles. Navigators use 1 minute of latitude as a useful unit called a **nautical mile**. It is 1.852 km. A speed of 1 nautical mile per hour is 1 knot.

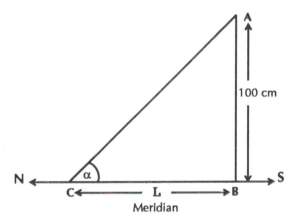

Fig. 1.17.1. The principle of measuring the radius of the Earth. ACB = Altitude of Sun α, measured by shadlow length L, than ACB = 100/L. A_S for Southampton, A_N for Newcastle.

The fraction $(A_S - A_N)/360$ is a measure of the ratio of the distance between Southampton and Newcastle to the circumference of the Earth $2\pi R$ (where **R** is the Radius of the Earth) which is assumed to be circular.

So

$$\frac{A_S - A_N}{360} = \frac{D}{2\pi R}$$

where D is the distace between the schools and can be measured on a good map using dividers and is found to be 452 km.

Thus the Radius of the Earth,

$$R = \frac{452 \times 360}{(A_S - A_N)\, 2\,\pi} \text{ km .}$$

The value of $A_S - A_N$ does not depend on the date, but dates in the spring or autumn will provide altitudes of the Sun convenient to measure.

For accuracy it is advisable for each group to spread three or four observations over a period of time, but keeping results tabulated so that the values A_S and A_N for corresponding dates are correctly subtracted. A Mean Value for R can be found.

Note: We can carry out this exercise to find R from a measurement of D from a map, but R is a well established quantity, the Earth's polar radius $= 6.3568 \times 10^3$ km.

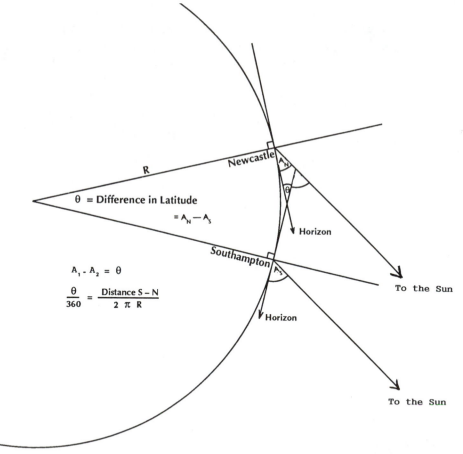

Fig. 1.17.2. Observations of the Sun's altitude at midday made at two schools separated by about 400 km can be made, and used to find the radius of the Earth, provided the two schools are on the same longitude. The diagram shows the exercise using a school at Southampton and another in Newcastle.

So to vary our exercise, we can accept R as 6.3568×10^3 km and calculate the distance D between the schools or cities which is the distance as "the crow flies" (502 km).

1.18   A DIY MODEL OF THE CELESTIAL SPHERE

Sky Watchers are generally familiar with diagrams of the celestial sphere given in books on astronomy, but for many these diagrams are not easy to visualise or understand. They provide little scope for things to do or for dealing with the basic terms used and how these are related and used in making practical observations.

The three dimensional model of the celestial sphere shown for latitude 51° in the Fig. 1.18.1 can be made easily and cheaply in a school or home workshop.

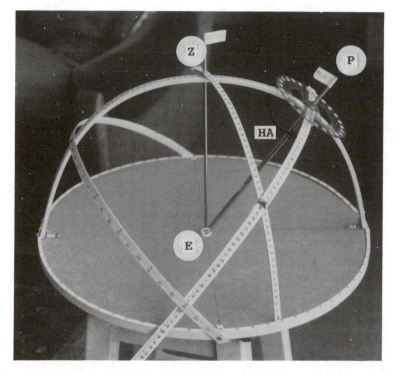

Fig. 1.18.1. A DIY model of the Celestial sphere. A 360° protractor is here used as a setting circle, marked in Sidereal Time Hours from 0h to 24h. EP is the polar axis parallel to the Earth's axis inclined at an angle ϕ = latitude of the Observer. EZ is the zenith tube,

$$\text{Length of tubes} = \frac{180}{2\pi} = 28.65 \text{ cm}$$

The circumference of the base circle = 180 cm.

The model can be used as a useful learning aid, as well as an instrument for star finding or identification since it shows clearly approximate values of the basic quantities involved. It can be used to give Sideral Time or Local Mean Time, and using the Sun it can indicate its Local Hour Angle and so it can be used as a sundial.

The circular base shown in the Figure is 180 cm in circumference, cut from a piece of hardboard 20 mm thick. The semicircular arcs of the various great circles as well as the edging round the base are made from standard strips of white plastic 10 mm wide and 2 mm thick commonly used in household repairs and furnishings. This scale is convenient as it ensures that all the angles shown by the strips, namely Altitude, Azimuth, Hour Angle and Declination, can be accurately marked using one centimetre to represent two degrees. Strips of centrimetric graph paper stuck on the plastic strips facilitates the graduation of the strips as shown. In the Fig. 1.18.1 ZE is a thin tapped vertical rod representing the overhead or Zenith position. EP is a similar rod representing the direction of the Earth's axis. Both rods are held securely in the positions shown, by small nuts to suit the threads of the rods.

Fig. 1.18.2. The DIY model of the Celestial sphere equipped with a sighting tube.

The model can demonstrate and explain all the usual astronomical terms used in positional astronomy, and illustrates the important relation used by the RA setting circle of telescopes, Local Hour Angle = Local Sideral Time − RA. The path of a star of known declination can be traced as it crosses the celestial sphere, and how this path depends on the declination of the star is clearly shown, including points on the horizon, of its rising and setting. A small piece of 'blue tac' placed on the declination circle at the correct declination can simulate any star, planet or comet. When set for the Sideral Time and Right Ascension the model will indicate the body's position in altitude, azimuth and Hour Angle. The use of a thin tube as shown in Fig. 1.18.2 connecting the observer's eye (E) and the blue tac will point convincingly to the body. The model, as mentioned, illustrates the operation of the equatorial sundial. The declination strip is turned about the polar axis till the strip casts a shadow on E. The time or the Sun's Hour Angle is read from the equatorial strip. A pencil positioned along the declination strip so that its shadow falls on E will show the Sun's declination and so show how the declination changes with the date. It also shows the Azimuth of the rising and setting of any celestial body, i.e. when its altitude is 0° and so the Azimuth is given by

$$\cos Az = \frac{\sin \delta}{\cos \varphi}$$

The model can be used to verify the results given by the Alt. Az. Dec. LHA nomogram, and the formulae of spherical trigonometry.

Fig. 1.18.3. The model shows a star at Altitude 42°, Azimuth 115°, Declination 20°, Hour Angle 3^h10^m or 47.5°. These values are in agreement with the values given by the nomogram, Fig. 3.7.2.

1.19   A 3D STARFINDER AND IDENTIFIER

Many have difficulty in relating a star's postion as seen in the sky with its position as depicted on a star map having the usual coordinates based on Right Ascensions and Declinations. Observers are generally advised to face a particular direction and hold the map or planisphere overhead, illuminated by a torch. This manoeuvre is not always a success, and can become uncomfortable. The device shown in Fig. 1.19.1 shows skywatchers precisely where to look on a specified date and time for any particular star given its Right Ascension and declination. To use the device, first find the Sidereal Time from the nomogram Fig. 1.13.5 by aligning the date and Local Mean Time, then position the hemisphere so that the RA mark corresponding to the local Sidereal Time is on the meridian, that is in the vertical plane passing through the North celestial pole and the centre of the hemisphere as in Fig. 1.19.1. The Fig. 1.9.2 shows for example the setting of the hemisphere for the 21st April at local time 23^h and sidereal time 13^h and also shows the star having RA 10^h08^m and declination 12°N. In the line of sight from eye to E to Star (Regulus). The converse procedure can of course give the RA and dec. of a star from a cursor sighting on a particular date and local time. So an unknown star or celestial object can be given its correct celestial position in terms of its RA and declination.

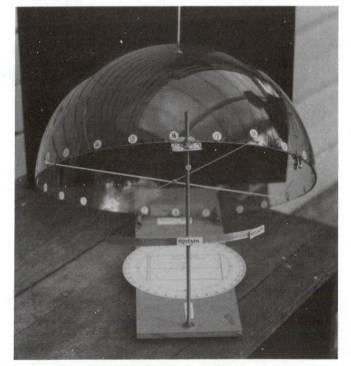

Fig. 1.19.1. Shows a hollow clear transparent plastic hemisphere about 150 mm in diameter, graduated in 24 divisions of Right Ascension. The cursor is a narrow strip of polythene graduated as shown in degrees of declination from 9° to 90° i.e. from the equator to the celestial pole.

It will be appreciated conversely that if we know the RA and declination of a star, then we can deduce the sidereal time by setting the device with the line of sight on the star and then noting the sidereal time that is in the meridian plane. Form this sidereal time and the date we can read from Fig. 1.19.2 the Local Mean Time. The device can thus be used to tell the time by observing one known star. It will be noticed that Fig. 1.19.1 shows the equatorial rim of the hemisphere marked suitably for observing stars in the Northern hemisphere of the celestial Sphere; but it can be instantly adapted for observing stars in the Southern celestial hemisphere, simply by reversing the polar axis EP to point to the South Celestial pole. It will thus still be parallel to the Earth's axis, and inclined to the horizontal at an angle equal to the Latitude of the observer. The RA gradations will however run round the equatorial rim in the reverse direction, i.e. clock wise in the Southern hemisphere, but anticlockwise in the Northern heisphere. Only about 25% of visible stars are in the Southern Hemisphere for observers in latitudes around 50°N.

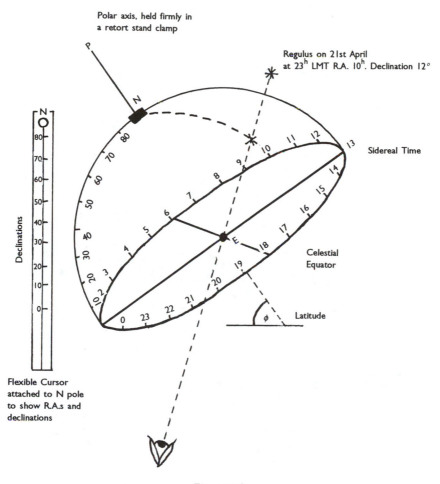

Fig. 1.19.2.

The device can be used to demonstrate the working of an equatorial sundial, by setting the hemisphere with its 12^h mark in the meridian. The cursor is then positioned so that its line casts a shadow on the central bead E. (Use a small slip of paper to receive the shadow), then the Hour Angle of the Sun (Sun Time) is the angle round the equatorial rim between the cursor and the 12^h mark in the meridian.

Users will be able to devise for themselves alternative ways for setting the device, as any conveniently placed bright star of known Right Ascension and declination can be used for this purpose. When set the hemisphere will instantly show the Sidereal Time, by the R.A. on the meridian, and can be checked by the sidereal time given by the nomogram Fig. 1.13.4 or from astronomical tables.

An additional instructive demonstration can be effected by placing the hemispherical rim on a horizontal plane surface so that the Right Ascension hourly gradations (at 15°) intervals, when suitably renumbered clockwise, then measure Azimuth angles. These will run from 0° (N) to 90° (E) to 180° (S) to 270° (W). The declination cursor now is in the position to measure altitudes up from the horizon toward the Zenith. Thus a useful distinction is made between the heavenly coordinates of RA and declination of the celestial sphere, and the Earthly coordinates based on our own particular place on Earth and our own Earthly horizon.

The device at present is for "Do it yourself" enthusiasts. Clear transparent plastic spheres and hemispheres are avaiable from firms dealing in decorations, light fittings and toys and can be adapted with a little workshop ingenuity to produce a 3D starfinder which can serve as a useful learning aid.

Fig. 1.19.3. The 3D Starfinder being demonstrated at a meeting of the British Astronomical Association.

2

The Sun and Sundials

2.1 THE SUN AS A TIMEKEEPER

The Pole Star was selected as a useful star to look at and study because it appears to have a steady and permanent place in the sky. It is thus a useful reference point in the Northern Hemisphere.

By contrast, our nearest star, the Sun, as seen from the Earth appears to follow a complicated path with a number of seemingly irregular habits. By making allowances for these small irregularities and averaging them out, it is possible to use the Sun to give us our accurate daily time, or Mean Time. Although the Earth spins at a fairly constant rate with respect to the stars, it also travels round the Sun once a year in an elliptical orbit with its axis tilted to make an angle of 66°.5 with the plane of its orbit, the ecliptic. These conditions cause our Sun time or Sundial time to be about 16 minutes ahead of the Mean Time clock around November, and to be behind Mean Time by about 14 minutes in February with smaller variations in between. The corrections to be made to Sun Time to get correct Mean Time or Universal Time are shown in the graph in Fig. 2.10.1 to be discussed later, under The Equation of Time. Throughout the ages, man has spent much time and technical skill in observing the Sun and his 'little sister', the Moon, for the important purposes of time keeping and for making calendars for regulating our communal activities, essential for the development and survival of organised life in communities, such as hunting, agriculture, migration and the organisation of religious and social functions. The ability to predict the phases of the Moon and eclipses played an important part in calendar making, and astronomers able to make these predictions held positions of great responsibility and respect. The Sun's movements in relation to those of the stars have from ancient times given us calendars, clocks and a sense of direction.

2.1.1 The solar system

Our small corner of the universe is aptly called the solar system as it is dominated and regulated by the Sun, which contains 99 percent of the total mass of the whole system which comprises the Sun, nine planets, their satellites, the asteroids and comets.

The experiments of Galileo in physics demonstrated the connection between masses, forces and velocities and helped to prepare the way for Kepler and Newton to formulate the laws of motion and to calculate planetary orbits. Newton was able to quantify the force of attraction F between any two masses M and m, separated by a distance d as being proportional to Mm/d^2, or $F = G\, Mm/d^2$, where G is a constant known as the Gravitational Constant. Using SI units with M and m in kg, d in metres and F in newtons, G the constant has the value 6.67×10^{-11} kg^{-1} m^{3}s^{-2}.

We all learn to respect gravity as soon as we begin to walk, as the most important force in the universe, and to appreciate its all pervading influence in ball games, athletics, rock climbing and astronomy. We live in a right angled world with the vertical (or plumb-line) given by gravity and the horizontal given by the horizon, which Pythagoras connected by his famous theorem about the square on the hypotenuse, the right angle which led to the development of trigonometry, vide section Fig. 1.10.4.

Newton was able to show that the force of gravity between the Sun and any of the planets was exactly the force necessary to keep it in its own elliptical orbit in accordance with Kepler's Laws. This force constantly pulls the planet in towards the Sun and so keeps the planet on its course otherwise it would fly off at a tangent.

2.1.2   How the Sun keeps us in order

Here is something to do. Suppose you wish to sling a small weight of 250 g on the end of a heaving line of twine over a tall tree or building. One way is to whirl it round on the string in a circle of about 1 metre radius in a vertical plane at about four times per second. This will give the weight a velocity round its circle 1m radius, of $2\pi \times 1 \times 4$ m per second, which is 25 m sec^{-1} and is faster than most people could throw it by hand (90 km/hr). At this speed it could reach a height of about 30 m if you let go of the string at the appropriate moment on its upward path.

At this speed of 25 m sec^{-1}-1 on a circular path the tension in the string, i.e., the pull on your hand is

$$\frac{mv^2}{r} = \frac{1}{4}\frac{25^2}{1} = 156 \text{ newtons}$$

or about 15.6 kg weight (1 kg weight is about 10 newtons); and to keep the weight in its circular path at this speed your hand would have to maintain a pull on the string of 15.6 kg weight or 156 newtons. In the solar system this is the kind of centripetal force that the Sun exerts on a planet, although the mysterious force of gravity takes the place of the tension in the string. A school physics course usually takes examples from bodies falling to the Earth under gravity or bodies being swung round on a string, but it is a change and more fun in this space age to calculate, for example, the force F that is keeping the Earth in its orbit round the Sun —

using the formula for the centripetal force on a body of mass m moving in a circle radius r, with velocity v,

$$F = \frac{mv^2}{r}, \text{ then}$$

using the values for m, v and r in Chapter 6 (6.6), this relation provides a good exercise in the use of SI units and the calculator giving

$$F = 3.5 \times 10^{22} \text{ N or } 3.5 \times 10^{18} \text{ tonnes.}$$

This is a force that a steel cable of about 5,000 kilometres thick could just withstand. This great force causes tides, earthquakes and weather changes.

We have used the relation

$$F = \frac{mv^2}{r}$$

for this force, but we could have obtained the same result using the Newtonian law of Universal Gravity which can be expressed as in 2.1.1

$$F = \frac{GMm}{r^2}$$

where M is the mass of the Sun and G is the Universal constant of gravitation. Equating the two values of F now obtained, we have

$$F = \frac{mv^2}{r} = \frac{GMm}{r^2}$$

and

$$G = \frac{rv^2}{M}$$

This is another exercise in dynamics with the calculator and is referred to in Chapter 5. The value of G so obtained agrees with the accepted value given in Section 2.1.1.

2.2   THE COPERNICUS REVOLUTION IN ASTRONOMY

The ancient stargazers had great difficulty in understanding and accounting for their careful detailed observations because they held firmly to the belief, based on the apparent fact, that the Sun, Moon , stars and planets all travelled round the Earth. Anyone who had the temerity not to acknowledge this self-evident fact and that the Earth was the centre of the celestial sphere, was either ridiculed or persecuted for holding heretical views that gave the world, its peoples and all creation anything less than a dominant position at the centre of all things.

Towards the end of the 15th Century astronomers experienced a revolution in outlook which began with Copernicus (1473–1543) who propounded a theory of planetary orbits of perfect circles round the Sun, and introduced epicycles to account for the retrograde motions of some planets. By the time of Galileo and Kepler, about 1600 AD, thanks to the invention of the telescope, the heliocentric system became well established.

The telescope in its simplest form was first developed for use in astronomy by Galileo in 1609, who demonstrated its remarkable capabilities by revealing some of the secrets of the solar system that were invisible to the unaided eye. This invention played an important part in the astronomy revolution. For instance, Galileo discovered four of the tiny Moons going round Jupiter each with a regular periodic time, which he demonstrated to astonished sky watchers. He observed that the shadows on our Moon were caused by mountains, and the inner planet Venus was seen to have appropriate phases as it orbited the Sun. Venus, and Jupiter with its Moons, were known to go round the Sun but Earth is a planet with a very fine Moon, so this was an additional argument for our Earth being a planet orbiting the Sun as other planets.

There is a well known hymn written by George Herbert (1535–1632) who was a young poet when Galileo's telescope was creating great interest among star gazers, philosophers and military observers all over Europe. Herbert was clearly excited by the discoveries made by looking through glass at the night sky and wrote this verse as a ''thought for the day''.

> *"A man who looks on glass (telescope)*
> *On it may stay his eye,*
> *Or if he pleaseth, through it pass*
> *And then the heavens espy."*

Stargazers following this time of revolution in outlook were overwhelmed with things to do, making, mounting and using telescopes with simple workshop equip-ment and applying Newtonian mathematics, to orbits, compiling tables using Kepler's Laws, and the calculus.

2.3 SUN SPOTS

Periodically the Sun's face breaks out into small spots which provide several interesting things to do using a telescope — *not* as an instrument to look through at the Sun, but as a means of producing a large projected image of the Sun, that clearly shows the spots which can be photographed, as in Fig. 2.3.1. These sunspots were noticed but not understood long before telescopes were in use as the unaided eye can see the spots under certain conditions such as at sunrise or sunset when the atmosphere acts as an effective filter to make unaided eye observation possible and safe.

Align the telescope on the Sun and allow the small image of the Sun formed by the eyepiece to fall on a piece of paper. Note that the Sun's image will burn a hole in the paper. **(This is a good demonstration of the danger that can result in blindness, of putting your eye anywhere near that hotspot)** — (see Section 4.10. *Photographing the Sun by eyepiece projection*). Now, hold a sheet of white cardboard about 20 or 30 cm away so as to catch the blurred image of the Sun and then move the eyepiece 1 or 2 mm in, to focus clearly the enlarged projected image on to the card.

Any spots on the Sun will appear as little dark patches a few mm across on the image with quite irregular shapes. To study these spots it is necessary to put a black collar round the telescope objective, and arrange for the Sun's image, which should be about 150 mm in diameter, to be received in a cardboard box, blacked inside. This size is convenient as 'Sun spot' discs can be obtained with co-ordinates marked on them so that the spot can be marked in pencil and be given the Sun's latitude and longitude for reference and comparison with other observers. Sun spots are the only readily available features of the Sun's surface that can be captured with a humble telescope. Their movements on the disc can be recorded over a period of days or weeks. These recordings with times and dates give us an idea of the Sun's rotational period (about 25 days), the movements of the Sun's photosphere, and the Sun's annual aspect.

2.3.2   Aurora and magnetic storms

Sun Spots are well worth studying, as they tell us a great deal about our nearest star and also about the millions of other stars that happen to be in the same modest class as our Sun, determined by mass, density and temperature. Further we can make our observations conveniently in daylight. By photographing the Sun's disc by projection, as in Section 4.11, Sun spots appear to move across the disc in about 12 days. Part of this movement is due to the rotation of the Sun, and part due to the Sun's surface being gaseous and moving with a different angular velocity from the Sun's interior. The spots appear to move more slowly near the poles of the Sun, and take about 17 days to travel half way round. The surface of the Sun is a raging inferno of very hot gaseous material called the **photosphere**, that is at a temperature of about 6000°C. The Sunspots we can observe are not the hot spots, they are darker than the rest of the Sun's surface, and are actually cooler than the photosphere having been blanketed by flares of gas that are blown out hundreds of thousands of kilometres high. Sunspots arise from violent explosions on the Sun's surface, that not only send out masses of gaseous material, but also streams of atomic particles known as The Solar Wind, and electromagnetic radiations. These on reaching the Earth produce the Zodiacal Lights aurora, and magnetic storms in our ionosphere that interfere with our radio and television, and can effect the Earth's magnetic field for short intervals of time. These changes can be detected using a simple home-made magnetometer. Figs 2.3.2 and 2.3.3.

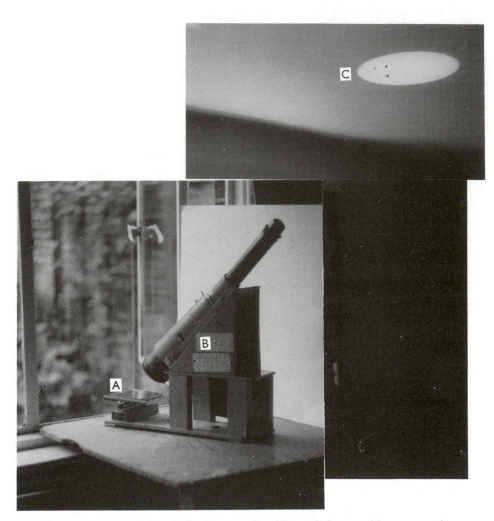

Fig. 2.3.1. A convenient and effective way of studying and photographing sunspots is to use as a small observatory, a room which has a south facing window. The Sun is reflected by a good top surface mirror A near the window sill, into a small telescope B which forms an image C on the white ceiling by eyepiece projection as shown in Fig. 2.3.1 and Fig. 2.3.2. The telescope is mounted on a firm stand, fixed at an angle of 50°, but the mirror able to turn on a horizontal axis to suit the altitude of the Sun, as described in Section 4.15 under the heading *"Observing with ease and comfort"*. The swivel mirror device of 4.15 uses the principle of Fig. 1.10.5 with a small telescope replacing the sighting tube. The small platform shown supporting a telescope at an open window as in Fig. 2.3.1 can be a great convenience as it transforms a living room or bedroom into a improvised miniature observatory. It thus relieves the skywatcher of the inconveniences of venturing outdoors at night. The platform can be used for supporting steadily binoculars, small telescopes, or a camera for short exposures or for photographing star trails as shown in Fig. 4.15.4.

Fig. 2.3.2. A cake container magnetometer.

The photograph, Fig. 2.3.2 and Fig. 2.3.3 show a polythene ''Cake container'' about 160 mm high, and about 350 mm diameter. A is a short bar magnet suspended from B by means of a fine nylon suspension, in the magnetic meridian, and near one side of the cake container. Attached to the magnet is a small piece of plane mirror C as shown. Diagonally opposite is a small 2.5 v self focusing bulb D, as used in small hand torches, and which has a hemispherical lens built into it. The bulb is threaded into the side of the vessel (D) and when connected to a small 3v battery, it conveniently focuses light on to the mirror, which reflects the light on to a graduated screen which is marked round the container as shown at E. A small change in the direction in the magnetic field of 10° is amplified by the reflected light by a factor of 2, and can be measured on the calibrated screen strip E. A change in direction of the Earth's field of 1°, produces a movement of the light spot F on the scale of 12.2 mm.

A similar device has been used by some Sunspot watchers using a magnet suspended in a jam jar, but I have found that the distortion of light rays entering and leaving the curved glass sides is very unsatisfactory, further complicated by having to arrange and mount a suitable support for the light bulb and for a screen about half a metre outside the jar. F is a fine ink line on the mirror which shows clearly on the screen.

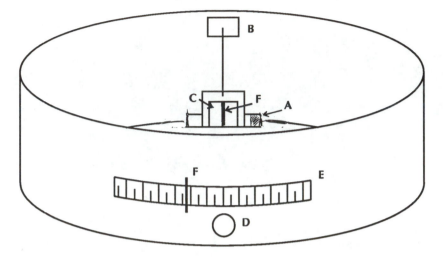

Fig. 2.3.3. A cake container magnetometer. A change in direction of the Earth's field of 1°, produces a movement of the light spot F on the scale of 12.2 mm given by

$$\frac{2}{57.3} \times 350 = 12.2 \text{ mm}$$

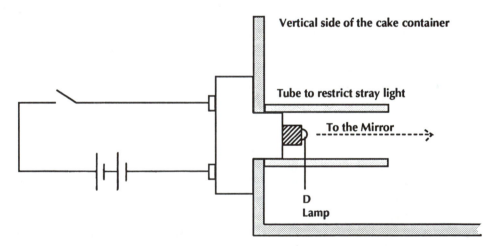

Fig. 2.3.4. Construction of the light source and sleeve in the cake container magnetometer.

2.4 SUNDIALS

The Sun in its apparent daily journeys across the sky gives us a means of telling the time and dividing our daylight hours into convenient intervals, by sundials. These Sun clocks throughout history, and in all parts of the world, have had a special fascination for astronomers, philosophers and country folk. Sundials were used in ancient Egypt and were often little more than sticks stuck in the sand, or vertical pillars. These were not at all accurate because the Sun varies its track across the sky according to the season of the year. A laboratory retort stand set up in a playground or garden will serve to show these variations from month to month by measuring the length and direction of the shadow it makes in sunlight. The vertical rod or retort stand can however be used to tell you both the azimuth of the Sun and its altitude, but we cannot mark on the ground the hour angle, or time by the Sun without tables and a mathematical calculation because this involves the declination of the Sun which changes with the seasons throughout the year. See Sections 2.8 and 2.9 for special vertical rod dials to solve this problem.

A great advance in accuracy and convenience was made by the Arabs who had the bright idea of tilting the rod in the plane of the meridian so that it pointed to the celestial pole and, this as we have seen, is at an angle with the ground equal to the latitude of the place, and consequently parallel to the Earth's axis. The part of a Sundial that casts a shadow on the dial is known as the style or **gnomon**. There are many types of Sundials but they can be grouped into three classes:

(1) Those having a gnomon or style parallel to the Earth's axis, and around which the Sun and all celestial bodies appear to move at the rate of 15° per hour. These dials are known as equatorial dials.
(2) Those that depend on the altitude of the Sun above the horizon, known as altitude dials.
(3) Those that depend on the Sun's azimuth or bearing known as **analemmatic** dials. These in general have a vertical style, which can be adjusted to suit the Sun's declination (Section 2.9).

2.4.1

The sundial having a style or gnomon inclined so that it points to the celestial pole is called an **equatorial** dial because the dial, that is the surface used to receive the shadow, is a flat disc in the plane of the celestial equator. The shadow angles so formed can however, be projected on to a horizontal plane or on to a vertical plane to make a horizontal dial or a vertical dial respectively. See Fig, 2.4.1.

Most books on Sundials written before the general use of calculators, provided complicated geometrical drawings and instructions for marking the shadow angles that are required for marking the hour lines on sundials, whether for a horizontal or a vertical dial.

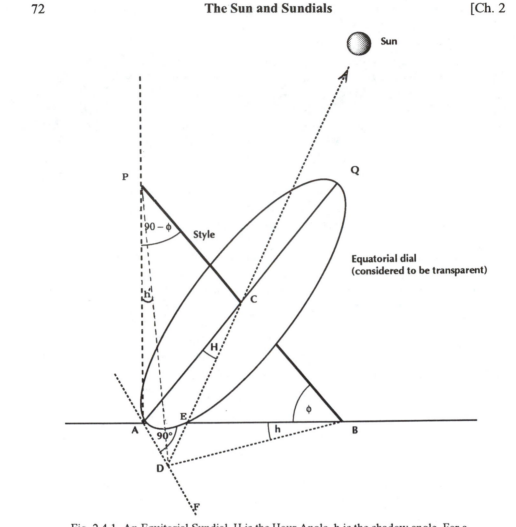

Fig. 2.4.1. An Equitorial Sundial. H is the Hour Angle, h is the shadow angle. For a
Horizontal dial tan h = tan H sin φ , for the Vertical dial tan h′ = tan H cos φ .

A little trigonometry as in Fig. 2.4.1 avoids the elaborate geometrical constructions, by using the simple relation which connects the Hour Angle of the Sun, (Sun
time) H, the angle h made by the shadow of the gnomon, and the latitude of the
place in which the Sundial is situated. The formula for a horizontal dial is:

$$\tan h = \tan H \sin \phi \qquad\qquad (1)$$

but for a vertical dial it is:

$$\tan h' = \tan H \cos \phi \qquad\qquad (2)$$

where φ is the latitude of the place. These relations are established in Fig. 2.4.1. In
this diagram, PB is the gnomom inclined at an angle PBA which is the latitude φ.
The Sun will cast a shadow CE on the circular disc which is on the equatorial plane

AEQ. This shadow makes an angle ACE on the disc and is the angle H the direction of the Sun S makes with the meridian line CA. The angle ACE = ACD is the Hour Angle of the Sun, and this is the angle that tells the time. The shadow CE meets the line AF at the point D. AF is in the horizontal plane and is perpendicular to AB the horizontal meridian line. ABD is the shadow angle h made by the style with AB the meridian line, on the horizontal plane.

From the diagram

$$\tan H = \frac{AD}{AC} \tag{1}$$

and

$$\tan h' = \frac{AD}{PA} \text{ and for a Vertical dial, } \tan h' = \frac{AB}{DA} \tag{2}$$

Now divide (2) by (1) we have

$$\frac{\tan h}{\tan H} = \frac{AD}{AB} \cdot \frac{AC}{AD} = \frac{AC}{AB} \text{ also } \frac{\tan h'}{\tan H} = \frac{AD}{PA} \cdot \frac{AC}{AD} = \frac{AC}{PA}$$

But

$$\frac{AC}{AB} = \operatorname{Sin} \phi \text{ and } \frac{AC}{PA} = \operatorname{Cos} \phi$$

therefore, $\tan h = \tan H \, \phi \sin \phi$ for a horizontal dial, and $\tan h' = \tan H \cos \phi$ for a vertical dial.

Table 2.4.1. Calculated shadow angles for a Horizontal Dial for Latitude 51° N. (London)

Sun time	Hour angle H	Shadow angle h
12	0°	0°00′
11/13	15°	11°45′
10/14	30°	24°10′
9/15	45°	37°51′
8/16	60°	53°23′
7/17	75°	71°00′
6/18	90°	90°00′
5/19	105°	109°00′
4/20	120°	126°32′
3/21	135°	142°09′

Formula: $\tan h = H \sin \phi$

Now give H the Hour Angles corresponding to the hours from 05^h am to 17^h pm and calculate the values of h (the shadow angles) for your latitude ϕ. The Table

2.4.1 shows the angles, which you can check, for latitude 51° (London). All you need now is a protractor and a sharp pencil.

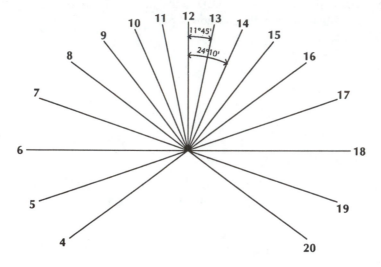

Fig. 2.4.2. Shadow angles for a horizontal sundial — Latitude 51°.

Fig. 2.4.3 shows a horizontal Sundial. If you wish to make a *vertical Sundial* to, hang on a south facing wall, the formula is similar but you will find that tan h′ = tan H cos φ as follows from Fig. 2.4.1, see also section 2.5.2.

Table 2.4.2. Calculated shadow angles for a Vertical South facing Dial for Latitude 51° N. (London) tan h′ = tan H cos φ

Sun time	Hour angle H	Shadow angle h′
12	0°	0°00°
11/13	15°	9°34°
10/14	30°	19°58°
9/15	45°	32°11°
8/16	60°	47°28°
7/17	75°	66°56°
6/18	90°	90°00°

Formula: tan h′ = tan H cos φ.

Fig. 2.4.3. A simple horizontal sundial made in a school workshop from aluminium sheet.

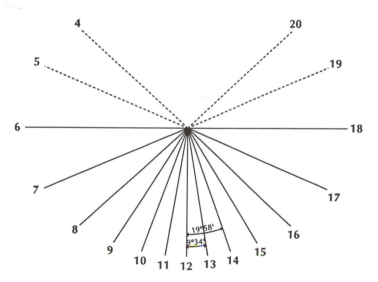

Fig. 2.4.4. Shadow angles for a south facing vertical sundial — Latitude 51°. See Fig. 2.4.5.

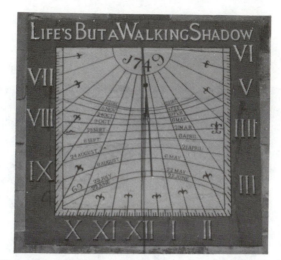

Fig. 2.4.5. This vertical dial has a notch on the gomon which casts a shadow on the dial, and marks the curved declination lines of the Sun ranging from Mid Summer, declination +23°.4, to the horizontal line for the equinox and to the short top curve for the Winter Solstice line with declination −23°.4. This sundial dated 1749 is in the Close of Salisbury Cathedral.

2.5 SUNDIAL DESIGNS

2.5.1 The Bicycle Wheel Sundial

A rewarding and instructive project for sundialists is to make a 'bicycle wheel' Sundial as shown in Fig. 2.5.1 in which AB is the gnomon 510 mm long pointing to the celestial pole and is a stainless steel rod 7 mm diameter. The shadow of this rod appears on a half of a cycle wheel. CD in the equatorial plane and is attached by a nut and bolt at E to the other half wheel AEF which is in the plane of the meridian. The two semicircles were obtained by cutting the rim of a cycle wheel into two halves. The gnomon, and the semicircle AEF are firmly attached to a piece of slate, S, approximately 420 mm × 200 mm × 15 mm. The photographs show the essential parts and how they are assembled.

It will be apparent that since the gnomon is parallel to the Earth's axis, the Sun's variable declination will cause the shadows of different parts of the gnomon to fall on the half rim CD. So in the summer, when the Sun's declination is +23°, the point T on the gnomon will fall on the graduated scale CD. In the spring and autumn, Sun's declination 0, the mid-point M of the gnomon will cast a shadow on the scale. In winter, the Sun's declination is −23°, and altitudes low, the point W will cast the shadow. The gnonom can thus, as a matter of interest, be graduated according to the declination of the Sun, in other words to the time of year. Marks corresponding to the declination can be scratched on the gnomon, and a small cursor ring round the gnomon or a pencil, can be moved along the gnomon till its shadow falls on the hour angle scale and can show the date to within a few days. A large bronze Sundial of great beauty by Henry Moore was originally erected in Printing House Square, London. This is an equatorial dial (as is the bicycle wheel dial) and the gnomon appears as a stretched bow string.

This pattern of a Sundial using a semicircular scale in the equatorial plane is regarded as the most accurate form in which a Sundial can be made and follows the form used by Jai Singh (1686–1743) for the great stone Sundial built in Jaipur in India.

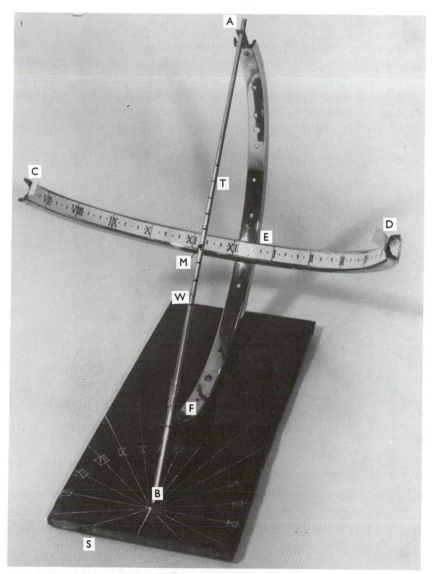

Fig. 2.5.1. The Bicycle Wheel Sundial. This model shows three projections of the stylus. E shows the projections on an equitorial disc also shown on the rim R. V shows projections on a vertical perspex sheet. H shows projections on a horizontal base plate.

In Fig. 2.5.1. the time scale CD is a strip of white plastic held to the equatorial rim, by clips or elastic bands, and which can be moved to the right or left to correct for the equation of time so that the XII mark appears on the Sun's 12 noon Mean Sun Time.

It will be seen in the photograph that the shadow of the style will fall on the XII mark about 16 minutes before the Sun crosses the meridian. In other words the Sun transits the meridian at $11^h 44^m$ Mean Sun Time and the equation of Time is consequentially 16 minutes. This value for E, the equation of Time, occurs in mid November as explained in section 2.9.

2.5.2 The two projections of the equatorial dial

The bicycle wheel sundial can conveniently be used to demonstrate these projections which are shown in Fig. 2.2.5. V is the vertical projection on to a sheet of perspex, E is the equatorial dial in the plane of the equator and H is the horizontal projection on to a slate base. These projections are described in Section 2.4.1.

Fig. 2.5.2. The Bicycle Wheel Sundial showing vertical, equatorial, and horizontal dials.

2.6 THE ALTITUDE SUNDIAL — HOW THE SUN'S ALTITUDE CAN TELL US THE TIME

The best known *altitude Sundial* is the portable dial which was a cylinder of wood illustrated in Fig. 2.6.1. The style is a horizontal projection across the top of the cylinder, the shadow of which shows the time, 10.30 am on May 4th.

It was known as a the Shepherd's or Traveller's dial and has a long history. It is attributed to a German monastic Scholar and mathematician of the early 11th Century.

Astronomy *can* give us points of interest in History and literature. In the *Canterbury Tales* by Geoffrey Chaucer (1340 to 1406) we are told that the ''gentle monk'' had a dinner date with the Good Wife of Bath and says to her:

> *''Goth now your way quoth he, all stille and soft, and let us dine as soon as that ye may, for by my chilindre it is prime of day.''*

The *chilindre* was his pillar dial, and was about 150 mm in height and 60 mm in diameter. It occurred to me that a standard 440 ml drinks can (e.g., a soft drinks can) might be used as the pillar to take the hour angle lines that can conveniently be inscribed on a sheet of A4 graph paper and wrapped round the can, see Fig. 2.6.1.

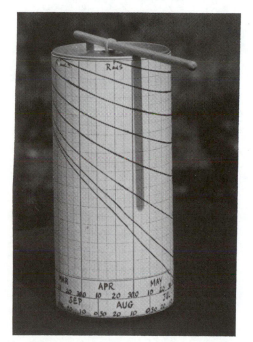

Fig. 2.6.1. A Pillar Dial (made from an empty drinks can!) showing Sun Time on 4th May as 10.30 am or 14.30 pm.

To make one of these historic dials, Fig. 2.6.3, the style should project radially a distance **d** from the edge of the can. The maximum length of the shadow is 130 mm when the Sun is at maximum altitude of 62.5° at noon midsummer, so that

$$\frac{130}{d} = \tan 62.5° \text{ and } d = 67.7 \text{ mm.}$$

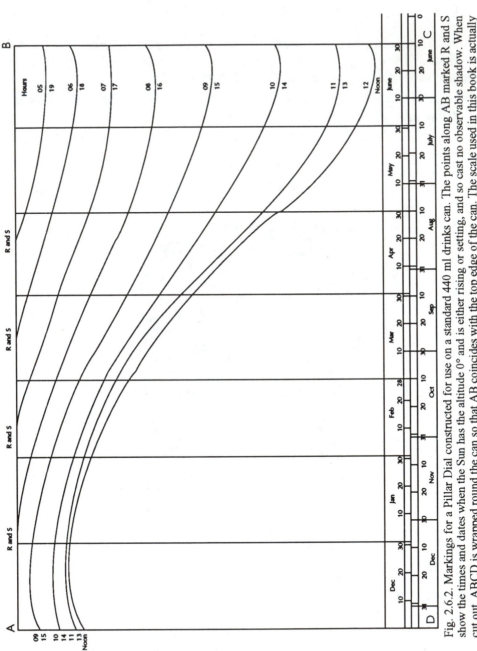

Fig. 2.6.2. Markings for a Pillar Dial constructed for use on a standard 440 ml drinks can. The points along AB marked R and S show the times and dates when the Sun has the altitude 0° and is either rising or setting, and so cast no observable shadow. When cut out, ABCD is wrapped round the can so that AB coincides with the top edge of the can. The scale used in this book is actually 75% of the actual size.

To plot the shadow hour lines from the dial to show how they vary with the Hour Angle of the Sun (HA), the altitude of the Sun (alt), the declination δ and the latitude ϕ of the observer, we use the spherical trigonometrical relation,

$$\text{Sin (alt)} = \text{Sin } \phi \sin \delta + \cos \phi \cos \delta \cos \text{HA}$$

non-mathematicians should not be put off by this, but use it as a programme for a calculator or computer to produce the Table 2.6.1 which contains all the data for plotting the set of graphs that is wrapped round the drinks can which are contained in Fig. 2.6.2.

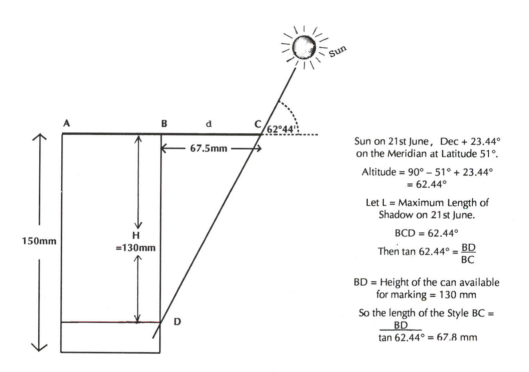

Sun on 21st June, Dec + 23.44° on the Meridian at Latitude 51°.

Altitude = 90° – 51° + 23.44°
= 62.44°

Let L = Maximum Length of Shadow on 21st June.

BCD = 62.44°

Then $\tan 62.44° = \dfrac{BD}{BC}$

BD = Height of the can available for marking = 130 mm

So the length of the Style BC =
$\dfrac{BD}{\tan 62.44°} = 67.8$ mm

Fig. 2.6.3. Calculating the length of the Gnomon on a Pillar Dial when the Sun is at maximum altitude of 62.44° at the Midsummer Solstice.

As an example, consider the altitude of the Sun on April 21st when its declination is 11.83°, in latitude 51° N and hour angle 60°, that is Sun Time = 8 am or 16 pm, 4 hours before or after midday.

$$\sin \text{(alt)} = \sin 51° \sin 11.83 + \cos 51° \cos 11.83° \cos 60$$

The calculator gives:

$$\text{(alt)} = 27.85° \text{ as in Table 2.6.1.}$$

Table 2.6.1. Pillar dial co-ordinates for curves (corresponding altitudes of the Sun appear in brackets) Latitude 51°

Date	Sun's declination in degrees	Times and Hour Angles							
		(Noon) 00	(11–13) 15	(10–14) 30	(9–15) 45	(8–16) 60	(7–17) 75	(6–18) 90	(5–19) 105
	x axis	*y* ordinates = 67.8 (alt)							
21st June	23.4	130.0 (62.45)	117.9 (60.1)	93.4 (54.015)	69.8 45.84	50.56 (36.71)	35.0 (27.29)	22.0 (18.00)	11.0 (9.19)
21st July 24th May	20.6	115.6 (59.6)	106.0 (57.4)	85.5 (51.59)	64.6 (43.62)	46.6 (34.49)	31.9 (25.21)	19.3 (15.86)	8.3 (6.95)
22nd Aug 21st April	11.83	83.1 (50.82)	77.9 (48.96)	65.1 (43.85)	50.2 (36.5)	35.8 (27.85)	22.8 (18.58)	10.9 (9.17)	0.0
23rd Sept 21st Mar	0	54.9 (39.0)	51.8 (37.44)	44.1 (33.02)	33.7 (26.42)	22.5 (18.34)	11.2 (9.37)	0.0	
24th Oct 18th Feb	−11.7	35.0 (27.29)	32.9 (25.95)	27.5 (22.09)	19.6 (16.15)	10.3 (8.64)	2.2 (1.826)	0.0	
22nd Nov 22nd Jan	−19.9	23.5 (19.09)	21.9 (17.87)	17.3 (14.35)	10.6 (8.85)	2.1 (1.80)	0.0		
21st Dec	−23.44	18.9 (15.54)	16.8 (13.94)	13.2 (11.00)	6.8 (5.7)	0.0			

2.7 A CALCULATION FROM THE 14th CENTURY

In the *Canterbury Tales* there are many references to astronomy and astrology which reflect the important part that mathematics and a knowledge of the spheres played in the cultural life of the 14th Century, for example, we read in the prologue to *The Parson's Tale*:

> "*The story of the Maniple had ended.*
> *From the south line the Sun had now descended So low, it stood — so far as I had sight —*
> *At less than twenty nine degrees in height*
> *Four of the clock it was, to make a guess: Eleven foot long, or little more or less,*
> *My shadow was, at that time and place Measuring feet by taking in this case*
> *My height as six, divided in like pattern Proportionately; and the power of Saturn Began to rise with Libra just as we*
> *Approached a little Thorpe....*"

(A Thorpe is a village)

It is interesting to note how Chaucer uses the altitude of the Sun, 29°, to give him the time, using the length of his own shadow of eleven feet, cast by his own height of six feet.

$$\text{Arc tan } \frac{6}{11} = 28.6°$$

The Sun is at this height at 4 pm when the Sun's declination is 12.5°. The Sun has this declination on about 22nd April.

This gives us the date. Chaucer could check this time using a Pillar Dial. On the pillar dial the shadow of the tip of the Gnomon (length 67.8 mm) would, when pointing in the direction of the Sun, be 37.75 mm from the top rim and the altitude of the Sun would be:

$$\text{arc tan } \frac{37.5}{67.8}$$

giving an altitude of 29.1°.

This kind of practical astronomy was part of an educated person's knowledge in Chaucer's day.

In Chapter 3 in the Section on Star Maps, Altitude and Azimuth Curves are used both to tell the time and also to obtain compass bearings graphically from the stars as well as from the Sun.

2.8   MAKING A VERTICAL STICK SUNDIAL

It is an instructive exercise to use a vertical stick to measure the altitude of the Sun and at the same time to measure its azimuth and then to use this altitude and azimuth to tell the time graphically.

In the diagram, Fig. 2.8.1, **AB** is a white vertical polythene tube 20 mm in diameter and 75 cm long fitted over a retort stand which is shown at the centre of a horizontal circle of 40 cm radius drawn on a level playground or, alternatively, on a square of hardboard as in Fig. 2.8.2. This circle NESW is marked in degrees of azimuth from 0 at the North point; the circle is oriented with its N–S line in the meridian. Attached to the rod is a sliding cursor **C** which can be moved up or down the tube which is graduated to read the altitude of the Sun when the shadow of the cursor falls exactly on the circumference of the circle.

Suppose the shadow of the cursor falls exactly at the point **D** on the Fig.. The altitude of the Sun is the angle **CDB** (α) given by

$$\tan \alpha \frac{CB}{40}$$

Now suppose that CB = 51 cm, then

$$\tan \alpha = \frac{51}{40}$$

and the altitude of the Sun is 52°.

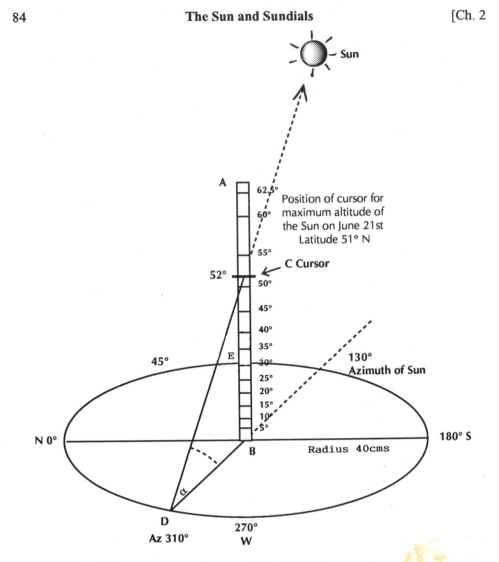

Fig. 2.8.1. A Vertical Stick Sundial being used to measure both the Sun's Altitude and Azimuth.

A table, Table 2.8.1, can now be drawn up to give the positions of C for the range of angles for the Sun's altitude which in latitude 51° is 62.5° which makes the maximum height of C to be:

40 tan 62.5 = 76.83 cm.

The positions of C for the range of altitudes of the Sun are given in Table 2.8.1 for latitude $\phi = 51°$. For intermediate altitudes the cursor reading is indicated by interpolation. A great advantage of this vertical style graduated so as to give the Sun's altitude, is that it also gives the Sun's azimuth as the 'compass rose' circle is graduated in degrees. In Fig. 2.8.1 the azimuth equals 310°. The style thus provides both altitudes and azimuths which can be entered together on a nomogram or a system of intersection curves to give the Sun's Local Time (Local Hour Angle) and the Sun's declination as will be described in Chapter 3. This makes it unnecessary to know the Sun's declination or the date.

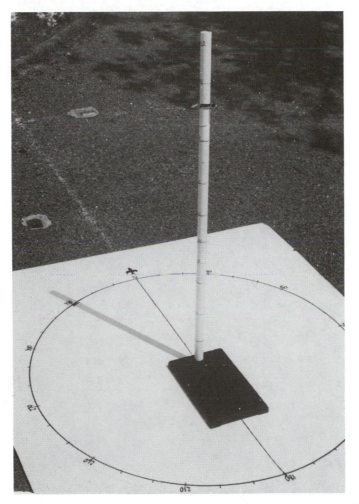

Fig. 2.8.2. The cursor A is moved along the vertical style, so that its shadow falls on the Azimuth Circle. Both altitude and azimuth are recorded by the shadow.

In Fig. 2.8.1 the azimuth of the shadow is 310°, so the azimuth of the Sun is 310° – 180° = 130° as shown. The altitude of the Sun is 52°. Knowing the altitude and azimuth of the Sun and the latitude it is possible to calculate the declination and the Hour Angle or Sun Time using spherical trigonometrical formulae

$$\text{Sin } \delta = \text{Sin } \phi \text{ sin alt} + \cos \phi \cos \text{alt} \cos \text{az for } \delta$$

and using values above for Az, Alt, and dec.

$$\text{Sin HA} = \frac{\text{Sin Az Cos Alt}}{\text{Cos } \delta} \text{ giving HA} = 30°.34$$

This is a little tedious even with a calculator, so in Chapter 3 Fig. 3.7.3 a graphical method is described which gives the result as HA = 30° (i.e., the Sun Time is either 2 pm or 10 am) and the declination is 21 as revealed at the point of intersections of the altitude curve 52° with the azimuth curve 130°.

Table 2.8.1. Positions of **C** in Fig. 2.8.1 for the
range of angles for the Sun's altitude in latitude 51°

Sun's Altitude α	40 tan α = BC
5°	3.5 cm
10°	6.7 cm
15°	10.7 cm
20°	14.6 cm
25°	18.7 cm
30°	23.1 cm
35°	28.0 cm
40°	33.6 cm
45°	40.0 cm
50°	47.7 cm
55°	57.1 cm
60°	69.3 cm
62.5°	76.8 cm

2.9   THE AZIMUTH DIAL

This dial, as the name implies, belongs to the third class of Sundials, which is the class that depends mainly on the compass bearing (azimuth) of the Sun. In Section 2.4 it is shown that a vertical style set up at the centre of a horizontal circle cannot properly show Hour Angles or tell the time throughout the year. From Fig. 2.4.1 we saw that hour angle lines can be correctly marked at 15° intervals for each hour, only if they are marked round an equatorial circle. So to form the horizontal base for an azimuth dial, a circle of radius (a) is divided into 24 hour lines at intervals of 15°. Imagine this circle to be held in the equatorial plane and projected on to the horizontal plane to form an ellipse with major axis 2 a and minor axis 2a sin ϕ, where ϕ is the latitude, as shown in Fig. 2.9.1. The minor axis must be in the me-ridian due N–S. It will be seen that the projected positions of the hour lines drawn from the centre of the ellipse are not equally spaced as they were on the equatorial circle. The Sundial angle γ measured from the N point is given by

$$\tan \gamma = \frac{\tan HA}{\sin \phi}$$

(which can be derived from Section 5.8.1 (3) and putting declination = 0°. These angles are marked on Fig. 2.9.1

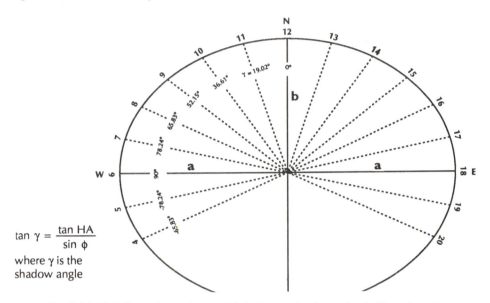

$$\tan \gamma = \frac{\tan HA}{\sin \phi}$$

where γ is the shadow angle

Fig. 2.9.1. This figure shows the sundial shadow angles drawn for the Sun when at declination 0°. The points at which these shadow angles cross the ellipse mark the hours of the dial. For other declinations the vertical style is positioned as shown in Fig. 2.9.2

$$\frac{b}{a} = \sin \phi.$$

The eccentricity of the ellipse is cos φ.

The construction and marking of this ellipse provides good scope for the calculator and understanding the properties of the ellipse. The dial is fun to mark out on a playground or garden and requires nothing more than some twine, a tape measure and a small pot of white emulsion paint, and can be carried out conveniently in a short time by a small team of three or four helpers. One of the attractions of this playground dial of the dimensions shown in Fig. 2.9.3 is that a young person, or slim adult, can stand at the appropriate spot on the meridian line and tell the time by the position indicated by his or her shadow, which falls on the periphery of the ellipse. (See Fig. 2.9.5.)

The actual playground dial Fig. 2.9.1 can be constructed by a group of 11 year olds, but the mathematics of the ellipse and the declination positions of the style will suitably occupy mathematics students.

2.9.1   How to mark out a playground dial

Choose a level patch of ground about metres square, and follow the instructions, using the dimensions shown on the plan of the layout, Fig. 2.9.3, with major axis of length 2a, (4 m) and minor axis in the N–S meridian, of length 2b (3.11m). The *meridian line* can be marked on a sunny day by placing a retort stand on the centre of the patch of playground being used, and noting and marking with a string the direction of the shadow when the Sun is exactly on your meridian. This time is given for the date in *Whitaker's Almanac* for the Greenwich meridian, so we must add 4 minutes for every degree you are west of Greenwich or subtract 4 minutes for each degree you are east of Greenwich. (If you are in Bristol about Longitude 2.5° west, you add $2.5 \times 4 = 10$ minutes.)

Fig. 2.9.3 also gives convenient dimensions for a playground Sundial, but for the purpose of drawing a plan on a sheet of graph paper, the angles are as in Table 2.9.1.

Table 2.9.1. The position of the shadow angles from the centre from:

$$\tan (\text{shadow angle}) = \frac{\tan(\text{HA})}{\sin\phi} \tag{1}$$

Sun Time	Hour angle	Shadow Angle
12	0°	0°
13/11	15°	19.02°
14/10	30°	36.61°
15/09	45°	52.15°
16/08	60°	65.83°
17/07	75°	78.24°
18/06	90°	90.00°
19/05	105°	−78.24°

2.9.2   Fixing the declination marks

In Figs 2.9.2 and 2.9.2(a), TR represents the vertical style of a playground Sundial, R is the centre point of the ellipse of the dial which has a semi-major axis CE of length 'a', and which as shown in Fig. 2.9.1, is the horizontal projection of the circular equatorial dial of which BP is the imaginary style, and which is parallel to the Earth's axis. The vertical style TR in this position will give correct hour angles for the Sun only when the declination of the Sun is zero. When the vertical style is moved into the position for the Sun having a declination of δ (about 20°) and the angle S°E S represents this declination.

We have seen in Section 2.5.1 that the shadow of the segment of the style at J will fall on the equatorial rim of the bicycle wheel dial, since J is in the path of the Sun's rays for this declination.

So in order that the vertical style will correctly record the same hour angle as the point J, the vertical style must be moved along the meridian BA of the horizontal dial a distance CF or R R° = D. From the figure CF/CJ = cos ϕ also CJ/a = tan δ so D = a tan δ Cos ϕ which is the relation for marking the distances D on the playground dial.

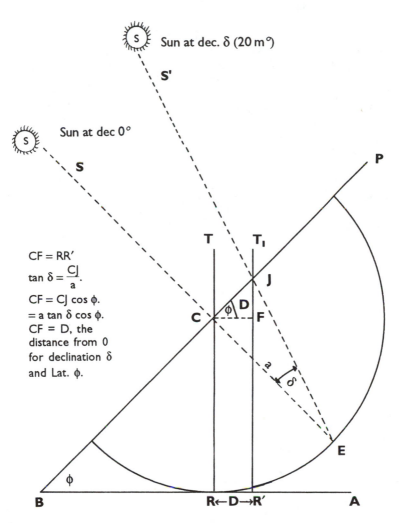

Fig. 2.9.2.

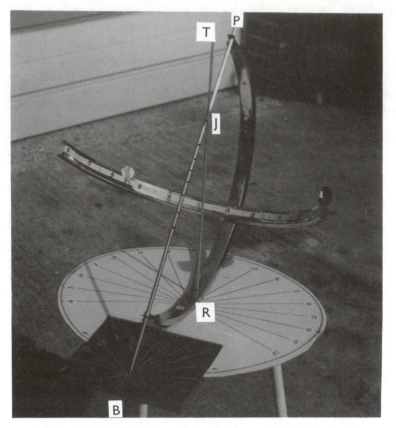

Fig. 2.9.2(a) Illustrates the geometry of Fig. 2.9.2.

Table 2.9.2. Position of the style for various declinations
from centre

$D = 100 \tan \delta \cos \phi$	(2)

Declinations	Distance of style from O
± 0	0
± 5	± 5.50 mm
± 10	± 11.10 mm
± 15	± 16.96 mm
± 20	± 22.90 mm
± 23.44	± 27.00 mm

These tables give only the main values required but a calculator can provide intermediate values for an accurate construction of this analemmatic dial.

The actual playground dial has a semi-major axis of 2 m, so the distances in Table 2.9.2 are 1/20 of their full scale.

Fig. 2.9.1 shows how the ellipse for the play-ground, or Azimuth dial, is formed by the projection of the equatorial circle on to a horizontal base. The hour angles marked on the circle at 15° intervals are not spaced at equal intervals on the ellipse.

2.9.3 Suggestions for marking out the ellipse for a playground Sundial

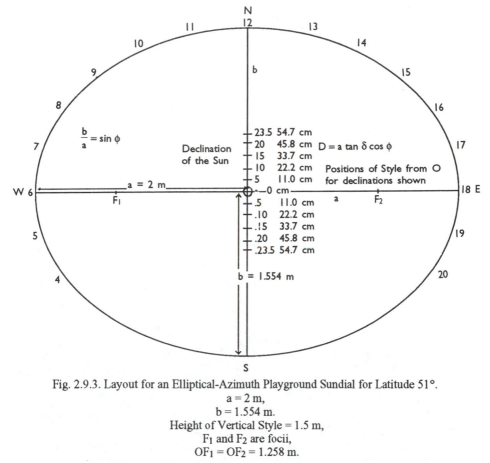

Fig. 2.9.3. Layout for an Elliptical-Azimuth Playground Sundial for Latitude 51°.
a = 2 m,
b = 1.554 m.
Height of Vertical Style = 1.5 m,
F_1 and F_2 are focii,
$OF_1 = OF_2 = 1.258$ m.

(1) Begin by marking the meridian line North–South in white emulsion paint making N–S = 3.11 m. We then use the well known property of an ellipse F_1 P F_2 is constant and = 2a = 4 m. Where P is any point on the ellipse.
(2) Mark the centre point O.
(3) Draw line WOE at right angles to N–S. This is the major axis of length 4 m.
(4) Mark points F_1 and F_2 at a distance of 1.258 m from O.

(5)   Secure a large nail at each end of a long string so that the distance between to two nails = 4 m.

(6)   Hammer one nail into the ground at F_1 and the other at F_2.

(7)   Stretch the string with its mid point at N and hold the string taut by a paint brush dipped in white emulsion paint.

(8)   Move the brush from N to P_2 and thus to P_1 to W. The path followed by the brush is part of the ellipse N P_2 P_1 W. Continue marking the ellipse till completed, keeping the string taut and the nails firmly in place. (See Fig. 2.9.4.)

(9)   Mark the shadow angles g radiating from the centre O as in Fig. 2.9.1 using a blackboard protractor, and the relation

$$\tan \gamma = \frac{\tan HA}{\sin \phi}$$

The angles thus marked are for the Sun with a declination of 0°, i.e., on March 21st or September 21st but give wrong results for other declinations. This limitation is overcome by moving the gnomon to the appropriate points on the minor axis as shown in Figs 2.9.3 and 2.9.5 corresponding to the Sun's declination. The distances δ from O along ON are given by

$$D = a \tan \delta \cos \phi$$

a calculation easily made on a calculator (see Table 2.9.2.)

   When δ is negative as in the winter, distances are negative, i.e., below 0.

Fig. 2.9.4. Marking out the sundial drawing the ellipse.

Fig. 2.9.5. Marking the declination points.

Fig. 2.9.6. Human style for the Sundial.

The photograph Fig. 2.9.5 shows a boy standing upright telling the time by using his head as a gnomon!

The declination marks at distances for the playground layout are shown in the figure. These distances which can be easily drawn on a sheet of A4 graph paper are conveniently scaled down by the factor 1/20. The Sun's declination shown can be given for each of the dates by looking them up in *Whitaker's Almanac* or in the *British Astronomal Association's Handbook*.

Azimuth Sundials, such as the one described in this section in which the observer plays the part of the gnomon or stylus of the sundial, can be an attractive feature of a garden or playground. It could be an instructive side-show at the proposed new Visitors' Centre of Stonehenge which is astronomically associated with the daily and annual movements of the Sun. See also Fig. 2.17.5, Section 2.17.

2.10   SUNDIALS AND THE EQUATION OF TIME

We have seen that Sundial time does not keep pace perfectly with an accurate clock, as it gets ahead of the clock at some parts of the year, and lags behind during other parts of the year. To overcome these differences, we use a Mean Solar Day as a unit which is the mean of all solar days taken over a year, then each 'Mean' day is considered to consist of 24 equal hours of Mean Solar Time, which for all practical purposes corresponds to Universal Time, U.T. or Greenwich Mean Time. (See the equation of time graph which is shown in Fig. 2.10.1.)

Until atomic clocks with an accuracy of 1 part in 10^{13} set our time keeping standards, our time measurements were based on astronomical observations and referred to the Earth's period of rotation with respect to the stars. The trouble with this procedure was that it defeated attempts to discover any variations in the period of the Earth's rotation itself! A homely example of this is the factory manager before the days of radio time signals who used to set his watch each morning when he passed a jeweller's shop, by the impressive clock in the window. One day he met the jeweller and expressed his appreciation of the jeweller's time-keeping service, as the manager's watch was, of course, the factory standard. The jeweller was a little embarrassed when he admitted that he always regulated his window clock by the factory hooter!

The Earth travels round the Sun very nearly in a circular orbit. It is an elliptical orbit of very small eccentricity of about 0.0167 in the plane of the ecliptic, and is part of the dynamic system involving the Sun, Earth and Moon.

The meridian line marked in your garden or playground Fig. 1.10.3 can be used to observe and keep a record of the exact time of the Sun's transit across the meridian during the course of the year. The results can be plotted on a graph as in Fig. 2.10.1.

2.10.1   Graphs of the two components of the Equation of Time

The Earth spins on its axis which is inclined to the plane of its orbit (the ecliptic) at an angle of 23°.44. It is possible, taking into account firstly, the effect E_1 of the eccentricity of the Earth's orbit and secondly, the effect of the obliquity of

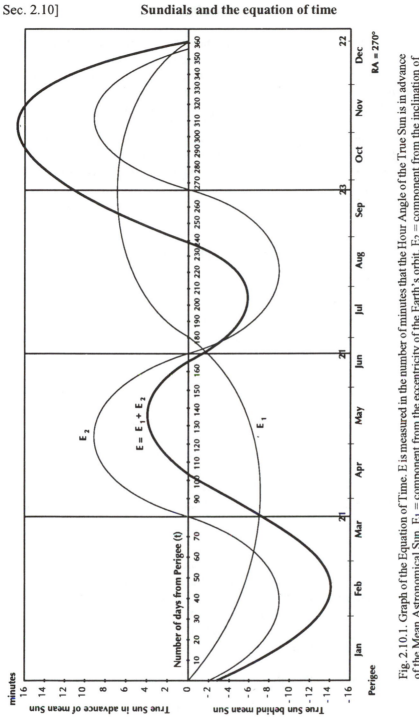

Fig. 2.10.1. Graph of the Equation of Time. E is measured in the number of minutes that the Hour Angle of the True Sun is in advance of the Mean Astronomical Sun. E_1 = component from the eccentricity of the Earth's orbit. E_2 = component from the inclination of the ecliptic, $E = E_1 + E_2$. (See also Section 5.13).

the ecliptic E_2, to calculate the combined effect $E = E_1 + E_2$ in order to account for the total differences E between True Sun Time (Sundial time) and Mean Sun Time (Universal Time). E is called the Equation of Time and is the correction to be applied to Sundial Time to give Mean Time as shown in Fig. 2.10.1 which shows the two separate graphs E_1 and E_2 and also the graph combining the two effects to produce E.

A method for forming the equations for plotting the graphs to show how the values for E_1 and E_2 (y axis) vary with the date throughout the year (x axis) is given in chapter 5 section 5.13 with the heading, ''An analysis of the Equation of Time''.

When E is positive, the True Sun is in advance of the Mean Sun and when E is negative, the True Sun is later than the Mean Sun. E is given in *Whitaker's Almanac* for each day of the year and is the correction to be applied to Sundial Time to give Local Mean Time.

Sundials often have inscribed on them a graph giving E for the year.

$$\text{Local Mean Time} = \text{Sundial Time} - E\,{}^{-\,\text{Long.W}}_{+\,\text{Long.W}}$$

For example, in early November E is 16 minutes. So a Sundial Time of 14.00 True Sun Time (observed in Bristol, which is 10 minutes West) is

$$\text{G.M.T.} = 1400 - 16 + 10 = 13^{\text{h}}54^{\text{m}}$$

or six minutes earlier than Greenwich Mean Time.

2.11   TO MEASURE THE ECCENTRICITY OF THE EARTH'S ORBIT ROUND THE SUN

An interesting exercise using a little mathematics and the calculator, is to calculate and measure the eccentricity of the Earth's orbit which is largely responsible for the Equation of Time. We can do this in fact, finding the ratio of the distance of the Sun from the Earth when maximum (apogee) to the distance when minimum (perigee).

This is $\dfrac{SD}{SB}$

from which e can be found (see Fig. 2.11.1). To measure the amount by which the Earth's elliptical orbit round the Sun deviates from a circle may seem a formidable task to carry out, but it requires only a simple calculation. The values for the semi-diameter of the Sun at apogee and perigee can be taken from *Whitaker's Almanac*, or the relative values could be found using a good telescope, and projecting an image of the Sun and measuring the maximum and minimum diameters of these images (see Section 4.11).

In Fig. 2.11.1, ABCD is the elliptical orbit of the Earth round the Sun as one focus.

S   is the Sun.
B   represents the Earth at perigee-when it is nearest to the Sun.
D   represents the Earth at apogee-when it is furthest from the Sun.

From B the Sun appears to have a semi diameter equal to $16'18''$ and from δ the Sun has a semi diameter equal to $15'43''$. The semi-diameters are in the inverse ratios of the distances between the Sun and the Earth, so,

$$\frac{SD}{SB} = \frac{16'18''}{15'45''} = \frac{KD+KS}{KD-KS},$$

but $KS = eKD$

$$\therefore \frac{16'.3}{15'.75} = \frac{KD(1+e)}{KD(1-e)} = \frac{1+e}{1-e}$$

and

$$e = \frac{16.3-15.75}{16.3-15.75} = 0.0176.$$

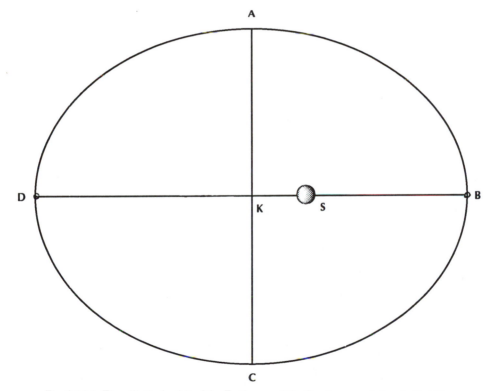

Fig. 2.11.1. The elliptical orbit of the Earth around the Sun (exaggerated as in fact S is much nearer to K)

2.11.1 Transits and Sidereal Time

In these days when most people have quartz watches, and know the G.M.T. accurately, it is easy to check the apparent time-keeping abilities of Sun and stars. All that is required is a well defined 'line-of-sight' along the meridian line (Az.180°). This, for example, can be fixed indoors using a South facing window, by marking on it a vertical line using a black felt-tipped marking pen such that an observer in the room uses this line as a foresight with a conveniently placed object such as a chimney or part of a building in the distance as the backsight of the meridian. For timing the Sun's transit of the meridian, use the shadow of the line and note the time when the shadow is due South, that is in line with the line and the fixed object. Compare your times with the transit times given in *Whitaker's Almanac* for the date in question. To allow for your longitude, add four minutes for each degree that you are West of Greenwich, or subtract four minutes for each degree that you are East of Greenwich. The device can be used to check that stars seem to be in a hurry to cross the meridian line and transit about four minutes earlier each night. This is borne out by reference to the nomogram illustrated in Fig. 1.13.5 which shows how star time (S.T.) depends on the date.

This homely transit device can be used on the planets as their times of transit are all given in the *Almanac*. Stars and their Right Ascensions, can often be identified by their transit times because at transit, the Sidereal Time is equal to the Right Ascension of the star.

A study of the Moon's behaviour can also be made using this transit device. Instead of hastening to cross the meridian each night, as is the case for stars, the Moon lags behind and appears on the meridian about 40 minutes later each night of its appearance, and even in daylight the Moon can often be seen so that a record of the Moon's transits can be kept for the whole of the month except for a few days when it is too near to the Sun to be observed.

There are, of course, accurate transit instruments with specially mounted telescopes aligned in the meridian, and adjustable in altitude, but a vertical line drawn on a South facing window pane is cheap, convenient and more comfortable to use, specially in midwinter.

The declination of the body is given by δ in the relation

$$\text{Altitude of Transit} = \text{Meridian Altitude of Equator} + \delta$$

The Altitude of the Equator is $(90 - \phi)$, or $\delta =$ the Altitude of Transit $- (90 - \phi)$. Thus for Altitude 17°, the declination is

$$17 \quad 17° - 39° = -22° - 39 = -22$$

and for Altitude 47°, the declination is

$$47° - 90° + 51° = +8° \, .$$

The device described in 1.10.5 can be used conveniently for observing transits in comfort.

2.12 NOON MARKS

The small irregularities of the Sun as a timekeeper can be easily recorded by means of a 'noon mark' which registers accurately the position of the Sun at 12 noon Local Mean Time. The Sun crosses your meridian, as we have seen in section 2.10 about 14 minutes late in February, then arrives 3 or 4 minutes early in May, then runs a little late in August and, in November, is about 16 minutes early.

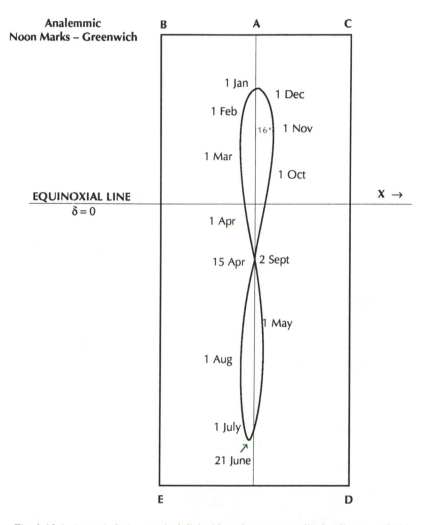

Fig. 2.12.1. A south facing vertical dial with nodus at perpendicular distance of 100 mm from the plane of the dial at Latitutde 51°N. When the Sun's image is on the curve this signifies that it is local noon (Mean Time), but this is later than G.M.T. noon by Longitude W or earlier by Longitude E.

All that is required for making a south facing noon mark is a vertical panel of white Formica 60 cm × 30 cm with a horizontal strip of metal projecting from the centre of the top edge. A central line is drawn with a "permanent" marker pen, and this line is the noon line upon which the shadow of the tip of the style or nodus will fall at the true Sun noon, i.e., when it is on the meridian. Our clocks, however, keep time by the Mean Sun Time so if we mark the shadow point, P, precisely each day at 12 Local Mean time (i.e., G.M.T. +Long. W or −Long. E) and note the date, at least once a week, and join the points, we get an interesting figure of eight, which is not quite symmetrical, such as that shown in Fig 2.12.1. Noon marks of this kind, but rather larger, were often set up on buildings and were used to correct clocks.

2.13   THE BLUE OF THE SKY

On clear days we all admire the blue of the sky but we may ask why and how is it blue? Here is something to do that attempts to answer these questions.

Without our atmosphere the sky would appear to be black, as experienced by astronauts when cruising outside our atmosphere.

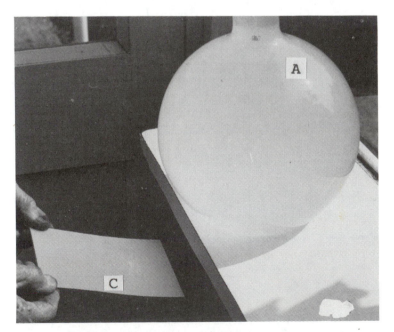

Fig. 2.13.1. Illustrating the blue of the sky. A reddish spot is formed on the screen by
light passing through the flask of cloudy water.

Fill a large spherical glass flask A, about 5 litres capacity, with water and add to it two or three drops of evaporated milk — just enough to make the water slightly cloudy. Hold the flask in the sunlight so that the Sun's rays fall on the flask, as shown in Fig. 2.13.1. The flask of water acts like a lens and will bring the incident light to a rough focus on a screen C, and the area round the focus will be distinctly reddish in colour. This reddish colour simulates the reddish hue of the rising or setting Sun. Sunlight, on encountering very fine particles in suspension becomes scattered, but the short wave blue light suffers more scattering than the longer red waves. The red light gets through with much less scattering than the blue light. The Rayleigh law of scattering states that the amount of scattering is inversely proportional to the 4th power of the wavelength. This means that blue light, which is about half the wavelength of red light, gets scattered 16 times more than red light. Blue light is thus dominant in light scattered from the sky. In addition, the blue scattered light is polarised. This polarisation of the blue light from the sky can be observed using a Polaroid spectacle lens, and in the same way can be detected in the light scattered from the side of the flask of cloudy water.

2.14   GETTING A BEARING OR CHECKING A COMPASS FROM THE SUN

Skywatchers and explorers can be led astray by the so called watch rule which states that due South can be found between the hour hand of a watch held horizontally with the hour hand pointing toward the Sun, and the 12 hour mark on the watch face. This simple idea seems plausible because the hour hand of the watch does move with an angular velocity twice as fast as that of the Sun: but what is ignored is the fact that the apparent path of the Sun and that of the hour hand of the watch are in quite different planes. This means that the hour angle of the Sun and the azimuth angle of the watch hand round the horizon are properly in coincidence only at 12 noon Local Sun time, and get more and more in error as time goes on.

Let us see what really happens for example at 3 pm on the 15th July when the Sun's declination is 20°N. The relation to use on a calculator is

$$\tan Az = \frac{\sin HA}{\cos HA \sin \phi - \cos \phi \tan \delta}$$

$$= \frac{\sin 45°}{\cos 45° \sin 51° - \cos 51° \tan 20°} \quad \text{(see Section 3)}$$

The ''Azimuth'' here is measured from the South point as in HA for true Azimuth add 180°.

This calculation gives the azimuth of the Sun as 245°.6, but the watch rule gives the azimuth as 225° which is an error of just over 22°. It would be quite disastrous if travellers were to use a compass that put them 22° off course.

Table 2.14.1. Azimuths of the Sun in degrees compared with bearings by the Watch method. Latitude 51°N, Bearings from N towards E

Local Sun Time	June +23°	May July +20°	April Aug +10°	Mar Sept 0°	Feb Oct −10°	Jan Nov −20°	Dec −23°	Watch Result
4	53							60°
5	64	66						75°
6	75		84	90				90°
7	86	88	95	102	108			105°
8	95	100	108	114	120	125	127	120°
9	112	114	122	128	133	138	139	135°
10	129	132	138	143	147	151	152	150°
11	152	154	158	161	163	165	166	165°
12	180	180	180	180	180	180	180	180°
13	208	206	202	199	197	195	194	195°
14	231	229	222	217	213	209	208	210°
15	248	246	238	232	227	220	221	225°
16	262	260	252	246	240	235	233	240°
17	274	272	265	258	252			255°
18	285	283	276	270				270°
19	296	294						285°
20	307							300°

The right-hand column shows the azimuth indicated if the 'Watch' setting is relied upon. Errors are a few degrees in mid-winter, but over 20° in the summer when the Sun is most likely to be used by travellers!

2.14.2   Risings and settings of the Sun

The Azimuth of any celestial body at rising or setting can readily be found from the spherical triangle formula $\sin \delta = \sin \phi \sin alt + \cos \phi \cos alt \cos A_z$ putting alt $= 0$ giving $\cos A_z = \sin \delta / \cos \phi$.

The way in which the Sun's Azimuth at rising and setting varies throughout the year on account of the way the Sun's declination varies can be shown by means of a circular diagram with the circle graduated in degrees of azimuth round the circumference from 0° to 360°, Fig. 2.14.3. This model can give the compass bearing of the Sun's rising and setting on any date of the year by interpolating between the dates marked, as necessary. The azimuth disc can be fitted with a sighting tube for observing azimuths with accuracy. It is of interest to make use of the fact that the actual path the Sun makes with the horizon as it rises from the horizon or sinks toward it is given by the angle θ which is given by $\cos\theta = \sin\phi/\cos\delta$. This angle is shown to vary only slightly from 39° (90°–ϕ) to 32.1° for $\delta = \pm 23.4°$.

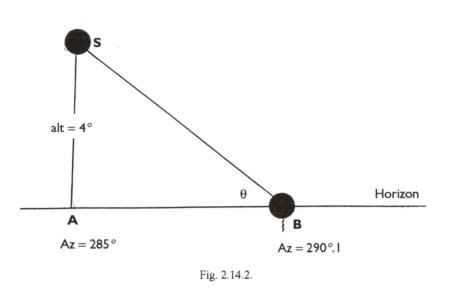

Fig. 2.14.2.

A knowledge of θ can be used for observing the actual azimuth of the Sun at rising or setting even though the Sun is a few degrees above the horizon, and it is perhaps impracticable to wait for the setting, which occurs when the centre of the Sun is on the horizon. For example, the Sun in Fig. 2.14.2, on 21st April is observed to be at an altitude of 4° above the horizon and at an azimuth of 285° as shown in the figure at A. The Sun will set at the point B and SBA = θ which from Fig. 2.14.2 is 38°. The azimuth of the setting Sun is thus 285° +AB but AB = SA/tan θ = 4/tan 38° = 5.1°. So the azimuth of the setting Sun is 290°.1.

The Fig. 2.14.3 shows on the circular disc, (reading from the periphery to the centre),

(1) The date,
(2) The Azimuth of the rising or setting,
(3) The angle at which the Sun approaches or recedes from the horizon,
(4) The Sun's declination, and
(5) The approximate times of rising and setting, using Time of rising = 12 – LHA, and the time of setting = 12 + LHA.

θ is the angle that the path of the Sun makes with the Horizon at the time of its rising or setting.

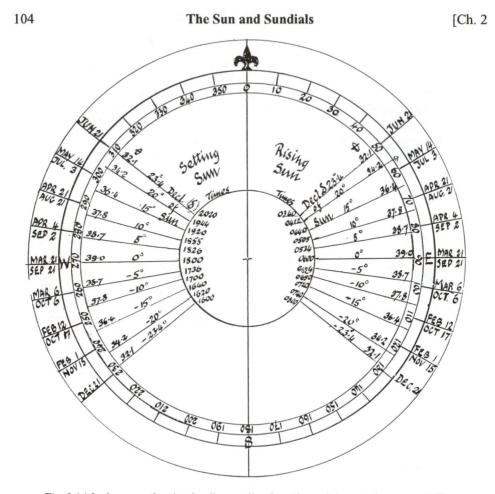

Fig. 2.14.3. shows on the circular disc, reading from the periphery to the centre, 1. The Date; 2. The Azimuth of the rising or setting; 3. The angle at which the Sun approaches or recedes from the horizon; 4, The Sun's declination, and 5. The approximate times of rising and setting.

2.15   UNUSUAL VARIETY OF AN EQUATORIAL SUNDIAL

The photographs illustrate an equatorial Sundial that may help readers to understand how the Sun appears to progress across the visible celestial hemisphere at different times of the year, and how the path followed depends on the Sun's declination and the date. This Sundial is unusual in that it has no fixed gnomon. The model shown consists of a clear transparent plastic hemisphere, 45 cms in diameter and marked in degrees of azimuth from 0° to 360° round the periphery, which represents the horizon, with its centre at C. Z is the zenith of the hemisphere, and CP is a thin tapped metal rod which is inclined to the horizontal at an angle φ, the latitude of the place. The celestial equator AB is marked in Hour Angles from 6^h (due East, Azimuth 90°) to 12^h (in the meridian plane, due South) to 18^h (due West, Azimuth 270°). The equator thus marked represents the path of the Sun when its declination is 0°, at the time of the Equinox. When the model is set as shown, the Sun's path during the Summer Solstice (declination 23°.4 degrees) begins at Sunrise at azimuth 50° E and ends with Sunset at azimuth 310° F. The Sun on this path crosses the meridian at an angle of 62.4 degrees in altitude above the horizon. During the Winter Solstice, (declination −23.4°) the path GH is from Sunrise at Azimuth 129° to Azimuth at Sunset Azimuth 231°. It reaches its maximum altitude at noon transit, of 15.6°. (39° − 23.4° = 15.6°), see Fig. 2.15.1.

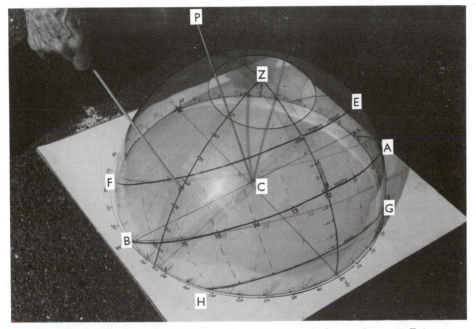

Fig. 2.15.1. The end of the knitting needle which casts a shadow on the centre E, is at the Sun's position. Time (H.A) 15^{h}40^m altitude 37° Azimuth 255°.

The Sun's position on the model can now readily be indicated by a knitting needle placed so that the shadow of the point falls on the centre C. The point on the model shows the Sun Time, and the declination of the Sun. The needle point is also directly above the Sun's azimuth marked on the horizon circle.

All the essential parameters of the Sun's position at any instant can be observed with a little interpolation between the markings. It can be an interesting and rewarding exercise to check these observed results by using the ''Sun'' nomogram shown in Fig. 3.7.3 which encapsulates the essential basic spherical trigonometrical formulae connecting the parameters.

For those who are familiar with the scientific calculator, it is satisfying to check the accuracy of the graphical results by using the two relations; (a is the altitude)

$\sin a = \sin \phi \sin \delta + \cos \phi \cos \delta \cos HA$ and $\sin \delta = \sin \phi \sin a + \cos \phi \cos a \cos H_Z$.

The nomogram demonstrates that it is possible to arrive at Sun Time from the Sun's altitude alone providing the Sun's declination is known. Similarly, the Sun's azimuth with the declination can provide the Sun Time. If however, both the altitude and the azimuth of the Sun are known, then the Sun's declination can be determined which yields both the date (approximately) and the time.

In order to facilitate the finding of the altitude and the azimuth of the Sun at any time a narrow strip of flexible plastic can be used as a cursor as shown in the photograph. This cursor is marked in degrees from 0° to 90°, that is from the horizon to the Zenith. The 90° point is pivoted by a small bolt at the Zenith Z. To facilitate the reading of the Sun's Hour angle (the Sun Time) a length of thread can be used, passing from the Pole point on the hemisphere, and extending to the Sun's position on the hemispheres. A little interpolation may be necessary to obtain an accurate time.

The use of the altitude-azimuth cursor together with the Hour Angle thread will convincingly demonstrate the danger that can befall beginners by confusing the hour Angle of the Sun with its azimuth. This danger can lead to serious errors in direction finding by the Sun, using the so called ''Watch Rule'' which is often erroneously advocated in books for outdoor adventure groups.

2.16   A PORTABLE POLAR SUNDIAL

The Polar Sundial usually consists of a rectangular plane dial surface that lies parallel to the Earth's polar axis, and has a central style fixed at right angles to the plane. The Hour Angle lines are drawn parallel to the Earth's axis and spaced at the distances x calculated by the relation $x = h \tan HA$, where h is the height of the style above the dial face, 6 cms. This relation results in x being inconveniently large for Hour Angles > 75°. For example at Sun times 7 am (HA = 75°), and 5 pm the dial would be 45 cms long, and to show times half an hour earlier and later than these times the dial has to be about one metre long. For the hours 6 am and 6 pm the dial would be infinitely long. $(x \tan 90° = \infty)$.

The photographs and figures 2.16.1, 2.16.2 and 2.16.3 show a simple model with hour lines marked using the style, edges BC and AD that are 6 cm high, and a length of dial of 24 cms that can accurately accommodate the full range of times from 6^h to 18^h. The dial is easy to make from a small sheet of aluminium ABCD 24×6 cms and 1 mm thick. The markings shown 2.16.1 can be made using a sharp scriber or ''permanent'' pen. The disadvantage of a metre long dial face is overcome by turning the two end squares AEFD and BCHG through 90° to form end pieces as shown, and also to serve as two styles, the edges AD and BC make the shadows when the dial is in position. The style BC will cast a shadow during the forenoon and show the hours marked from 6 am to 12 noon. The style edge AD then takes over the task and shows the afternoon hours from noon to 18^h. The shadow of BC just covers the edge AD at 6 am, and the shadow of AD just covers the edge of BC at 18^h as marked. In this way the problem of having an infinitely long dial face is overcome.

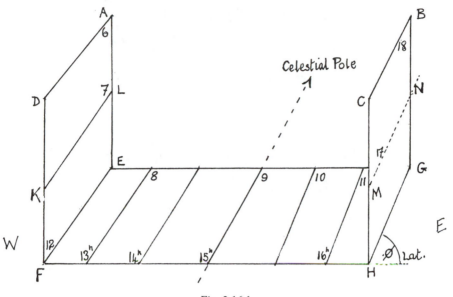

Fig. 2.16.1.

Instead of calculating the positions of the hour lines, the positions can be found graphically using a protractor and then marking the angles shown in Fig. 2.16.4. This also shows how the two end styles reduce the unwieldy length of the dial from "infinity" to 24 cms, as each style end undertakes the task of showing the hour angles for 17^h, 18^h and 7^h, 6^h. The half hour markings should not be made by dividing the intervals into two equal parts, but by calculation or by using half Hour Angles, in their constructions.

This Polar Dial can readily be adjusted in position to correct its readings for the Equation of time, or for its Longitude East or West. This is done by using a small wedge to raise the Western edge FE by a few mms which effectively adds a few minutes to the Local Sun time recorded, or correspondingly by raising the Eastern edge of the model the Local Sun times are reduced, and in each case the increase or decrease is at the rate of four minutes for each degree raised.

The wedge shown in the photograph 2.16.3 can be calibrated to facilitate these adjustments. As an example during early November, the Equation of Time or $4°$ ahead in our Angle. The correction for this can be made by simply turning the face FEHG through $4°$ about the edge FE, by raising HG by $FH \sin 4° = 12 \sin 4° = 8.3$ mm using the calibrated wedge P in the photograph. Similarly during mid February the equation of Time is −14 minutes so the wedge is inserted under the Western edge FE to raise it by 14/4 minutes that is by 12 sin 3.5 = 7.3 mm.

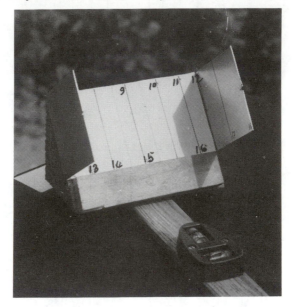

Fig. 2.16.2. Dial showing time 11^{h}40^m.

Fig. 2.16.3. Dial showing time 15h20m. P shows the correcting wedge (Not actually in use).

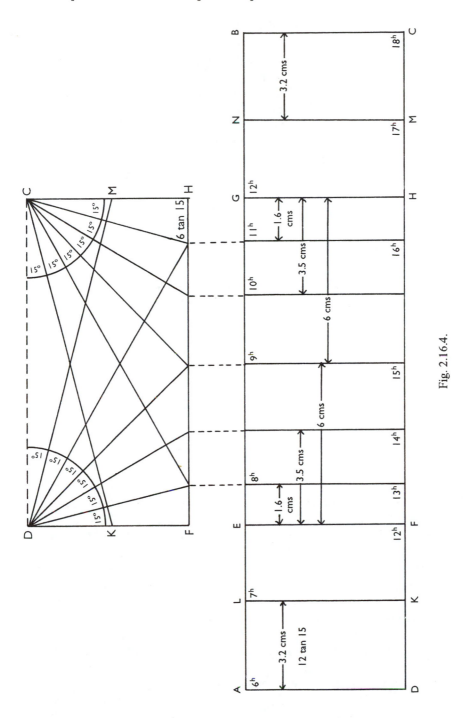

Fig. 2.16.4.

2.17 ECLIPSES FROM THE ASTRONOMY OF STONEHENGE

The people of Stonehenge were concerned with the worship of the Sun and Moon as were many other early civilizations. The Sun figured as a God and the Moon as a Goddess, so that eclipses of the Sun and Moon became events of great importance and communal excitement. Efforts to understand and predict eclipses occupied the attention of savants and leaders of their day. Anyone who could predict an eclipse could rise to a position of great eminence.

From recent studies of Stonehenge it became clear that Stonehenge was much more than a device for forming a calendar to mark the seasons and days of the year from a circle of stones. This was the practice on many monolithic sites.

An eclipse occurs when the Earth, Sun and Moon are perfectly in line. If the Earth lies between the Sun and Moon we have a lunar eclipse; if the Moon lies between the Sun and the Earth we have a solar eclipse.

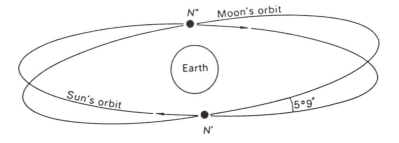

Fig. 2.17.1. The paths of the Sun and Moon on the sky lie in planes which intersect at an angle of 5′9′

As shown in Fig. 2.17.1 the apparent paths of the Sun and the Moon as seen from the Earth appear to intersect at the points marked N′ and N″ the nodes of the Moon's orbit. The Sun's apparent path in the sky is regular thoughout the year, but the Moon's path changes from day to day and from year to year in a complex manner, which Newton admitted gave him a headache. The main reason for this complexity is that the Moon's orbit round the Earth is inclined to the plane of the ecliptic at an angle of 5° which causes the angle between the extreme positions of the Moon's rising to vary from 60° to 100° as in the Fig. 2.17.4.

The early observers at Stonehenge found that it was possible to mark the positions of the Sun, Moon and the nodes and to keep track of them on a large circle of 87.7 m in diameter. This was done by dividing the circle into 56 equal parts by means of 56 holes (known as Aubrey holes) with hole number 1 to represent the Sun at the Summer solstice, Fig. 2.17.3. The number 56 must have been arrived at by a stroke of genius as would have come from a Newton or Galileo of 200 BC, as it elegantly enables the Sun to follow an annual path of 364 days, the Moon to go round the circle in 28 days and the nodes to complete the circuit in 18.6 years, by the following simple procedure, using markers to represent Sun, Moon and nodes, as suggested by Professor Sir Fred Hoyle, in his book *On Stonehenge*.

(1) The Sun marker moves anti-clockwise 2 holes every 13 days and so completes its circuit in one year of 364 days.

(2) The Moon marker also moves anti-clockwise, but 2 holes every day, and so completes the circuit in a lunar month of 28 days. There are 13 lunar months in a year.

(3) The nodes move clockwise, 3 holes every year, and so complete the circuit in 56/3 years (18.6 years) which is near the Saros period of 18 years 11 days known to the ancient Chaldean astronomers as the period between eclipses.

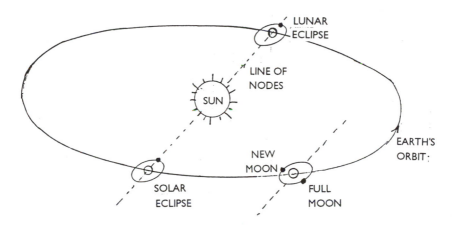

Fig. 2.17.2. Solar and Lunar eclipses. An eclipse, Sun or Moon occurs when the Sun, the Moon, the nodes and the observer at E on the Earth are in alignment.

If the markers are initially placed with the Sun, Moon and Nodes together to represent a total eclipse, then, by following the procedures 1, 2 and 3, the markers will again all be together for an eclipse after 18.6 years.

A total eclipse of the Sun observed from any particular place is a rare event. For example we had only two total solar eclipses in the British Isles in this 20th Century, one in 1927 and another due in 1999. Eclipses of the Moon however are much more frequent, because the Earth is much larger than the Moon, and its shadow is widely spread out, and so total or partial eclipses of the Moon occur once or twice each year, and could have been approximately predicted using near alignments of markers.

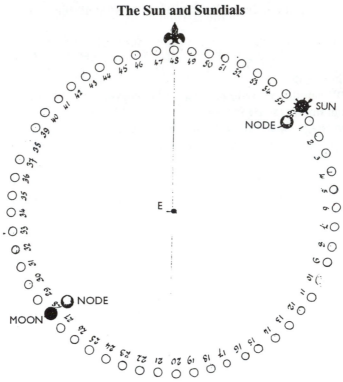

Fig. 2.17.3. An eclipse of the Sun or Moon occurs when the Sun, Moon, Earth and the Moons nodes are aligned with the observer on the Earth at E.

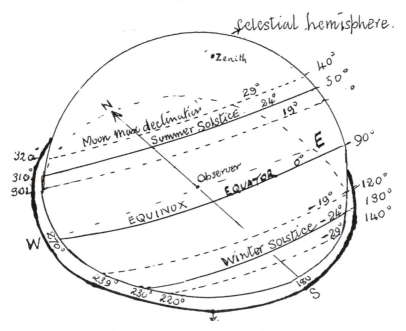

Fig. 2.17.4. The horizon or Azimuth circle.

Fig. 2.17.4 depicts the apparent paths followed by the Sun and the Moon as they move across our celestial hemisphere. The Sun is shown at its two extreme positions at the Summer and Winter solstices. The Moon's paths are shown by dotted lines and illustrate the effect of the rapid changes in the Moon's declinations over half a metonic cycle of 9.3 years, between the extreme positions given by ±(24° + 5°) and ±(24° − 5°). It is these swings of declination that account for the ranges of Azimuths of the risings mentioned above. Maximum range in the risings of the Moon is from Az. 40° to 140° and the minimum from 60° to 120°, i.e. Max range is 100° and Min. range 60°. These figures can readily be checked using a calculator on the simple relation,

$$\text{Cos Az} = \frac{\text{Sin } \delta}{\text{Cos } \phi},$$

where δ is the declination and ϕ is the latitude, (51° Stonehenge).

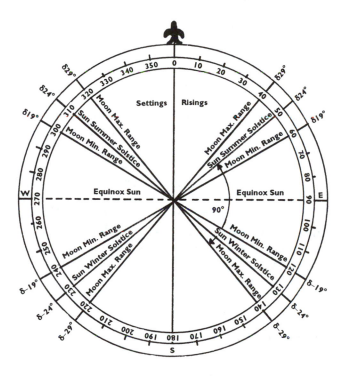

Fig. 2.17.5. Showing the azimuths of the risings and settings of the Sun throughout each year, and the wide montly variations in the Moon's risings and settings during the 18.6 yearly period required for the precision of the Moon's nodes to complete the circuit round the ecliptic. This period is the period between two solar eclipses and could have been studied and used in predicting eclipses by the buiders of Stonehenge.

3

Star positions, Star maps, Planispheres and Nomograms

3.1 THE NEED FOR SPHERICAL TRIGONOMETRY

The most realistic means of representing stars for sky watchers is the star globe. A star globe shows stars as we see them if we imagine ourselves looking outward from the centre of the celestial sphere. Stars have positions defined by Right Ascensions and Declinations as described in chapter 1 section 1.8. A convenient way of marking a star map on a flat surface is to use polar co-ordinates, that is with Right Ascension circles shown as straight lines radiating from the North celestial pole (for Northern Hemisphere maps). These lines are usually spaced at intervals of $15°$ or one hour. The declination circles are drawn at $10°$ intervals and crossing the Right Ascension lines orthogonally with declination $0°$ at the equator and ranging from the equator to the North Pole declination $+90°$ to declination $-90°$ at the South celestial pole, as in Fig. 3.1. A circular map of this nature is the basis of a planisphere and is designed to turn about the pole as an axis. The planisphere includes the stars of its hemisphere, but the stars visible at any particular time are revealed by an horizon mask that also can turn about the pole. This mask not only defines the observer's horizon for the observer's Latitude but also can indicate the times at which stars rise or set, and the Sidereal Times at which they cross the meridian.

3.2 FINDING AND IDENTIFYING STARS

We rely on maps and charts for finding our way about the Earth and the night sky. Before printing produced maps, as well as books, cheaply, and so changed our outlook on education, globes for use in geography and astronomy were in common use as learning aids and the ''use of the globes'' was generally taught as part of a sound school education. Widely separated positions of places on the Earth cannot be accurately depicted on a plane sheet of paper without distortion. Try wrapping a small sheet of paper round a football.

On a geographical map the distance between two towns a hundred miles or so apart can be found by measuring the distance between them but for long distances a straight line joining the places does not give the correct distance unless the map has been specially drawn. Navigators and air-line pilots are well aware of this.

Ordinary plane geometry and trigonometry have to be replaced by spherical trigonometry which deals with triangles on the surface of a sphere. The rules are similar to those of plane trigonometry and can be taken care of by a scientific calculator.

When we look at the clear night sky, we are presented with a picture of the celestial hemisphere which is turning slowly anti-clockwise about the North celestial Pole with our horizon as the boundary of the hemisphere. A star globe is a useful piece of equipment; when it is set for Sidereal Time, say T hours, and with its

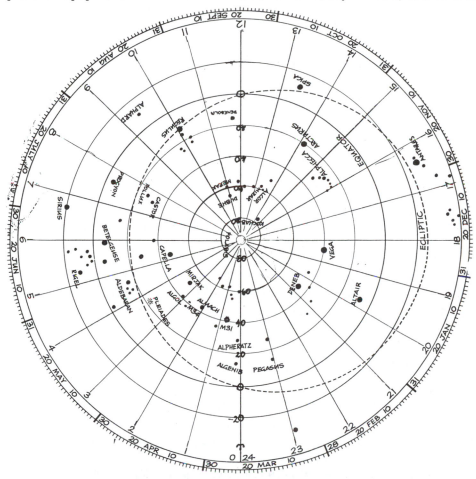

Fig. 3.1. A simple Planisphere Star Map. Showing RA lines and concentric declination polar circles. The date mark coincides with the Sun's transit (local 12 noon) is thus RA (Sun) + 12 ^{h}.

North pole pointing to the North celestial pole, then any star which is then in transit across the meridian has the RA of T hours. All stars at this instant are then correctly orientated with respect to the celestial sphere. As we watch a transit of a star as it appears to move westwards across the meridian, the angle the star's RA makes with the N–S meridian is called the star's Hour Angle (HA).

This is a very important quantity for telescope users and navigators. It can be expressed simply as,

HA = Sidereal Time − RA

and so measures how far round the celestial sphere the star has moved since its transit. It can be thought of as the angle between the RA of the star and the sidereal time. It will be appreciated that a star globe correctly positioned and used with a sighting tube will give a fairly accurate result for the altitude and bearing of any object that is, or can be, marked on the globe.

3.2.1   The limited usefulness of star maps and planisphers for observers
Star maps or planispheres with star positions depicted in terms of Right Ascensions and declinations do not indicate the all important position of any star in terms of our natural and commonly used position co-ordinates, namely its altitude above the horizon and its azimuth, or bearing, measured round the horizon from the North point (see Fig 3.2). These positions change continuously with time but can however be calculated accurately by a scientific calculator or better, by a computer programmed for the task to produce printouts for a particular latitude. As in Tables 3.3.1 and 3.3.2.

In order to use the planisphere star map, when it is set for the date and the local time, it is necessary to hold the map above your head, map downward with the midnight mark pointing toward the North. This enables you to relate the star positions on the map with the actual stars in the sky; but the method is somewhat awkward and especially inconvenient in the dark. A means of helping beginners to overcome this difficulty has been described in section 1.19, which explains the use of a hemispherical 3D starfinder.

3.3   COMPUTERS AND CALCULATORS

Calculator or computer print-outs for Latitude 51° (London) are shown in Tables 3.3.2 and 3.3.3. Tables for latitudes 40°N (for Southern Europe and the Middle States of the USA) and 57°N (for North Britain, actually for Edinburgh) are given in Appendix 6.4. Thus with the requisite interpolation, latitudes from 35° to 60° can be provided for with fair accuracy.

When the figures are plotted as co-ordinates on polar graph paper, they produce a pleasing and useful set of curves in the form of an *intersection nomogram* which relates the four parameters — altitude, azimuth, Hour Angle and declination — for a particular time and date, see Fig. 3.3.1 and Fig. 3.4.1.

The pattern of curves can be photocopied to the appropriate scale on to a sheet of good tracing paper, or acetate, and used as an overlay on a planisphere as shown in Fig. 3.4.2, and set for 19th August at 22^h00 LMT.

This transforms the planisphere into a versatile instrument with many uses such as:

- identifying stars,
- a star compass and a Sun compass
- times of risings and settings of stars and planets knowing their RAs and declinations.
- a Nocturnal for telling the time by the stars.

Further examples of the use of alt–az graticules are shown in Figs 3.4.3 and 3.4.4; these include the *Philip's Planisphere*, the *Daily Telegraph Planisphere* and the *Greenwich Star Disc*.

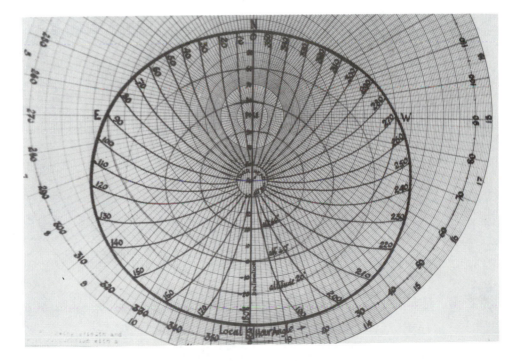

Fig. 3.3.1. Altitude and Azimuth lines for a Planisphere Overlay. Drawn from the computer printouts 3.3.2, 3.3.3 and 3.3.4 (Latitude 51°N). This figure shows the horizon which clearly defines the visible celestial sphere.

Table 3.3.1. Table of co-ordinates of points for plotting *altitude* curves or *almucantars* for use on star charts of various types, Polar, Cartesian, Mercator or Stereographic projection. The relation used is Sin (alt) = Sin φ Sin d + Cos φ Cos d Cos (HA) or as rearranged

$$\cos(\mathrm{HA}) = \frac{\sin(\mathrm{alt}) - \sin\phi\sin\delta}{\cos\phi\cos\delta} \quad \text{where } \phi \; 51°\mathrm{N \ or \ S}$$

Dec \ Alt	5	10	15	20	25	30	35	40	45	50	55	60	65	70	75	80	85
80	0.00	0.00	0.00	0.00	0.00	0.00	0.00	0.00	122.00	89.63	60.50	22.87	0.00	0.00	0.00	0.00	0.00
75	0.00	0.00	0.00	0.00	0.00	0.00	0.00	131.48	105.51	84.58	65.14	44.91	17.14	0.00	0.00	0.00	0.00
70	0.00	0.00	0.00	0.00	0.00	0.00	136.72	113.98	96.18	80.43	65.61	50.90	35.13	13.36	0.00	0.00	0.00
65	0.00	0.00	0.00	0.00	0.00	140.20	119.45	103.38	89.40	76.58	64.42	52.56	40.59	27.76	10.40	0.00	0.00
60	0.00	0.00	0.00	0.00	142.73	123.36	108.42	95.51	83.78	72.81	62.33	52.17	42.15	32.06	21.43	7.76	0.00
55	0.00	0.00	0.00	144.70	126.36	112.24	100.06	89.02	78.74	68.99	59.62	50.54	41.65	32.89	24.17	15.28	4.99
50	0.00	0.00	146.29	128.77	115.27	103.63	93.08	83.26	73.96	65.04	56.41	48.00	39.76	31.65	23.63	15.67	7.71
45	0.00	147.64	130.79	117.79	106.57	96.39	86.90	77.90	69.26	60.89	52.71	44.66	36.70	28.74	20.65	11.99	0.00
40	148.81	132.53	119.96	109.07	99.18	89.95	81.17	72.71	64.50	56.44	48.47	40.52	32.46	24.07	14.66	0.00	0.00
35	134.08	121.86	111.26	101.61	92.57	83.96	75.64	67.53	59.54	51.59	43.59	35.39	26.70	16.63	0.00	0.00	0.00
30	123.58	113.23	103.77	94.90	86.42	78.20	70.16	62.20	54.24	46.16	37.81	28.83	18.20	0.00	0.00	0.00	0.00
25	115.03	105.75	97.07	88.64	80.50	72.49	64.55	56.55	48.40	39.89	30.64	19.52	0.00	0.00	0.00	0.00	0.00
20	107.58	98.96	90.68	82.59	74.62	66.67	58.64	50.40	41.73	32.23	20.66	0.00	0.00	0.00	0.00	0.00	0.00
15	100.81	92.59	84.56	76.60	68.63	60.55	52.22	43.40	33.66	21.67	0.00	0.00	0.00	0.00	0.00	0.00	0.00
10	94.42	86.42	78.47	70.48	62.34	53.91	44.95	34.97	22.60	0.00	0.00	0.00	0.00	0.00	0.00	0.00	0.00
5	88.22	80.27	72.25	64.05	55.52	46.41	36.21	23.47	0.00	0.00	0.00	0.00	0.00	0.00	0.00	0.00	0.00
0	82.04	73.98	65.72	57.08	47.81	37.39	24.30	0.00	0.00	0.00	0.00	0.00	0.00	0.00	0.00	0.00	0.00
-5	75.70	67.35	58.61	49.19	38.54	25.10	0.00	0.00	0.00	0.00	0.00	0.00	0.00	0.00	0.00	0.00	0.00
-10	69.00	60.14	50.55	39.68	25.89	0.00	0.00	0.00	0.00	0.00	0.00	0.00	0.00	0.00	0.00	0.00	0.00
-15	61.69	51.94	40.83	26.68	0.00	0.00	0.00	0.00	0.00	0.00	0.00	0.00	0.00	0.00	0.00	0.00	0.00
-20	53.36	42.00	27.49	0.00	0.00	0.00	0.00	0.00	0.00	0.00	0.00	0.00	0.00	0.00	0.00	0.00	0.00
-25	43.23	28.32	0.00	0.00	0.00	0.00	0.00	0.00	0.00	0.00	0.00	0.00	0.00	0.00	0.00	0.00	0.00
-30	29.20	0.00	0.00	0.00	0.00	0.00	0.00	0.00	0.00	0.00	0.00	0.00	0.00	0.00	0.00	0.00	0.00
-35	0.00	0.00	0.00	0.00	0.00	0.00	0.00	0.00	0.00	0.00	0.00	0.00	0.00	0.00	0.00	0.00	0.00
-40	0.00	0.00	0.00	0.00	0.00	0.00	0.00	0.00	0.00	0.00	0.00	0.00	0.00	0.00	0.00	0.00	0.00

Table 3.3.2. Table of co-ordinates of points for plotting *azimuth* curves relating to latitutde 51°N for use on star charts of various types, Polar, Cartesian, Mercator or Stereographic projection. The relation used is:

$$\tan \delta = \frac{\sin \phi \cos (HA) + \sin (HA) \cot (Az)}{\cos \phi}$$

Az δ(HA)	10	20	30	40	50	60	70	80	90	100	110	120	130	140	150	160	170	180
10	70.22	63.14	59.45	57.09	55.36	53.98	52.78	51.67	50.57	49.42	48.13	46.58	44.56	41.58	36.44	24.61	-19.23	-90.00
20	76.74	69.35	64.56	61.05	58.26	55.85	53.64	51.48	49.25	46.79	43.91	40.25	35.16	27.15	12.36	-18.41	-62.51	-90.00
30	79.83	72.91	67.76	63.62	60.06	56.80	53.65	50.42	46.92	42.90	37.96	31.41	21.94	6.99	-17.05	-48.07	-73.78	-90.00
40	81.56	75.08	69.78	65.19	60.99	56.93	52.81	48.39	43.41	37.45	29.87	19.61	5.08	-15.18	-39.46	-61.74	-78.34	-90.00
50	82.60	76.41	70.99	65.98	61.15	56.25	51.04	45.24	38.44	30.08	19.33	5.20	-12.82	-33.30	-52.74	-68.59	-80.70	-90.00
60	83.23	77.19	71.57	66.11	60.56	54.69	48.20	40.70	31.69	20.55	6.65	-10.04	-28.25	-45.64	-60.48	-72.46	-82.08	-90.00
70	83.58	77.54	71.61	65.57	59.17	52.10	44.00	34.44	22.90	9.04	-6.91	-23.74	-39.71	-53.62	-65.20	-74.80	-82.92	-90.00
80	83.72	77.51	71.12	64.32	56.79	48.19	38.10	26.12	12.10	-3.52	-19.55	-34.57	-47.69	-58.79	-68.17	-76.24	-83.41	-90.00
90	83.67	77.10	70.03	62.16	53.13	42.53	30.04	15.65	0.00	-15.65	-30.04	-42.53	-53.13	-62.16	-70.03	-77.10	-83.67	-90.00
100	83.41	76.24	68.17	58.79	47.69	34.57	19.55	3.52	-12.10	-26.12	-38.10	-48.19	-56.79	-64.32	-71.12	-77.51	-83.72	-90.00
110	82.92	74.80	65.20	53.62	39.71	23.74	6.91	-9.04	-22.90	-34.44	-44.00	-52.10	-59.17	-65.57	-71.61	-77.54	-83.58	-90.00
120	82.08	72.46	60.48	45.64	28.25	10.04	-6.65	-20.55	-31.69	-40.70	-48.20	-54.69	-60.56	-66.11	-71.57	-77.19	-83.23	-90.00
130	80.70	68.59	52.74	33.30	12.82	-5.20	-19.33	-30.08	-38.44	-45.24	-51.04	-56.25	-61.15	-65.98	-70.99	-76.41	-82.60	-90.00
140	78.34	61.74	39.46	15.18	-5.08	-19.61	-29.87	-37.45	-43.41	-48.39	-52.81	-56.93	-60.99	-65.19	-69.78	-75.08	-81.56	-90.00
150	73.78	48.07	17.05	-6.99	-21.94	-31.41	-37.96	-42.90	-46.92	-50.42	-53.65	-56.80	-60.06	-63.62	-67.76	-72.91	-79.83	-90.00
160	62.51	18.41	-12.36	-27.15	-35.16	-40.25	-43.91	-46.79	-49.25	-51.48	-53.64	-55.85	-58.26	-61.05	-64.56	-69.35	-76.74	-90.00
170	19.23	-24.61	-36.44	-41.58	-44.56	-46.58	-48.13	-49.42	-50.57	-51.67	-52.78	-53.98	-55.36	-57.09	-59.45	-63.14	-70.22	-90.00
180	-51.00	-51.00	-51.00	-51.00	-51.00	-51.00	-51.00	-51.00	-51.00	-51.00	-51.00	-51.00	-51.00	-51.00	-51.00	-51.00	-51.00	-51.00

Table 3.3.4. The Horizon Curve

The co-ordinates for the altitude curve (alt) = 0 i.e., for the horizon are given by the altitude relation used for Table 3.3.1.

$$\cos(HA) = \frac{\sin(alt) - \sin \phi \sin \delta}{\cos \phi \cos \delta} \text{ and when } \sin(alt) = 0, \text{ then } \cos(HA) =$$

$$- \tan \phi \tan \delta$$

The calculator gives the following table from

$$\tan \delta = \frac{-\cos(HA)}{\tan \phi} \text{ with } \phi = 51 \text{ which makes the horizon curve easy to draw.}$$

HA			δ	
0	and	360	−39.00	
10	and	350	−38.57	
20	and	340	−37.27	
30	and	330	−35.04	
40	and	320	−31.81	
50	and	310	−27.50	
57.62	**and**	**302.38**	**−23.44**	**Winter Solstice**
60	and	300	−22.04	
70	and	290	−15.48	
80	and	280	−8.00	
90	and	270	0.00	
100	and	260	8.00	
110	and	250	15.48	
120	and	240	22.04	
122.37	**and**	**237.62**	**23.44**	**Summer Solstice**
130	and	230	27.50	
140	and	220	31.81	
150	and	210	35.04	
160	and	200	37.27	
170	and	190	38.51	
		180	39.00	

From this curve the times and azimuths of the risings and settings of all bodies having declinations between ±39 can be found. This horizon curve is approximately elliptical in shape. The HA for the Sun when corrected for longitude and equation of time will give the local time of Sunrise and Sunset. A polar type planisphere has to have the centre of its horizon ellipse (the zenith) at a distance of (90 − φ) from the poles where φ is the latitude of the place.

3.4   A NOTE ON PLANISPHERES AND HOW THEY CAN BE PUT TO MAXIMUM PRACTICAL USE USING AN ALT–AZ GRATICULE

There are three kinds of planisphere; they differ only in the co-ordinate system used for mapping the stars.

Type I is typified by *Philips' Planisphere* of which the star maps are circular with RA lines radiating as straight lines from the Pole at 15°, i.e., at hourly intervals. The declinations are circles concentric with the pole, equally spaced at intervals of 10° from the pole 90° to the equator 0° and about 40° beyond to –40°. The horizon is an oval as shown in Figs 3.4.1 and 3.4.2. Figs 3.4.3 and Fig. 3.4.4 show the Alt-Az graticule in use on various planispheres.

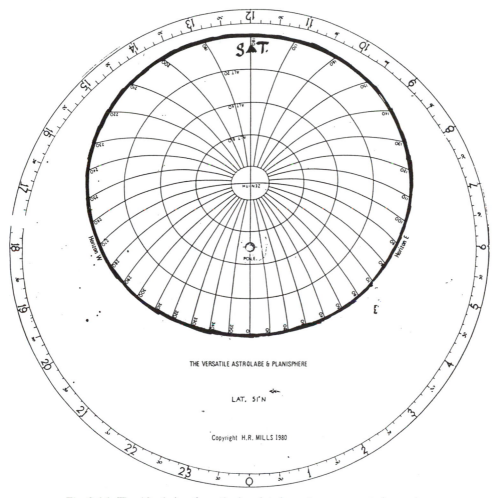

Fig. 3.4.1. The Alt.-Azimuth graticule printed on a transparency to be used as an overlay on a star map shown in Fig. 3.4.2.

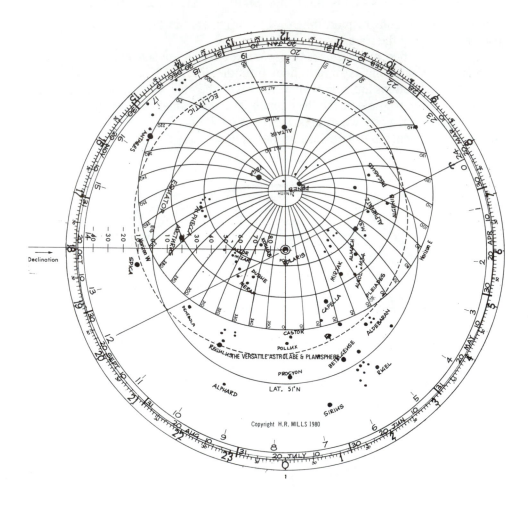

Fig. 3.4.2. This shows the transparency of the Alt.-Az. graticle applied as an overlay
on the star map to show star positions in terms of their Azimuths and Altitudes. For
example at 22^h00 on 19th August, the star Altair of the Summer Triangle will be seen
on the meridian, at an altitude of 48°. (The Local Mean Time 22^h, is brought into line
with the date, 19th August.) Arcturus is at alt 20° and Az 280°.

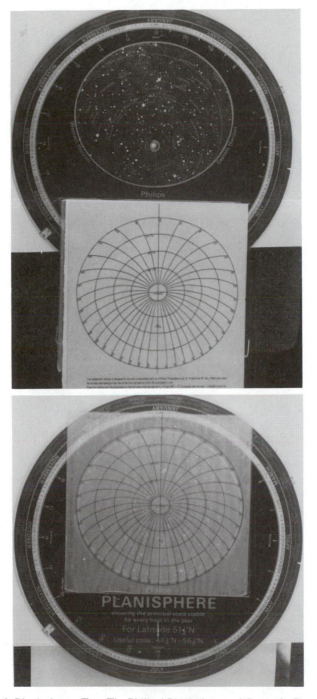

Fig. 3.4.3. Planispheres. Top: The Philips' Planisphere and Graticule. Bottom: The overlay in position over The Philips' Planisphere.

Fig. 3.4.4. Top: The Daily Telegraph Star Chart with its Alt-Az overlay. Bottom: The Greenwich Star Disc with its Alt-Az graticule. These two chargs are being turned by a 24 hour clock which keeps the charts satisfactorily on Sidereal Time.

3.5 TYPE II PLANISPHERES STEREOGRAHIC PROJECTION

This type of planisphere is in principle identical with the mediaeval astrolabe that for many centuries was by far the most important scientific intrument used by astronomers and navigators. It differs from Type I planispheres in one respect. The star map is formed by stereographic projection.

This unusual projection was a brilliant innovation by the ancient Greek mathematicians and astronomers who skilfully avoided spherical trigonometry which was at the time little known or practised. For this projection, the view of the heavens as seen from the South Pole is used as shown in Fig. 3.5.1. The projection being made on the equatorial plane. In this way the declination curves are all circles that become more and more spread out as they get nearer to the equator, in accordance with the relation

$$r_\delta = \frac{r_0 \tan (90 - \delta)}{2}$$

where r_δ is the radius of the circle for declination δ and r_0 is the radius at the equator.

The medieval map is as shown in Fig. 3.5.2, the graticule showing altitudes and azimuths is shown in Fig. 3.5.3.

The special feature of this astrolabe is that the altitude and azimuth curves are all circles which are orthogonal, that is to say that they all intersect each other at right angles. The astrolabe was the ingenious answer of the early Alexandrian astronomers to the problem of how best to represent on a map the positions of stars as they move across the sky, at a time when spherical trigonometry was practically unknown.

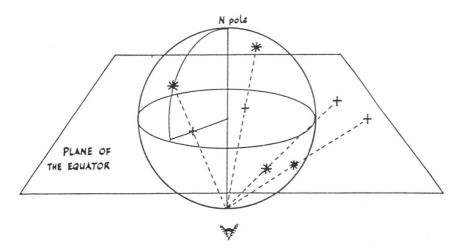

Fig. 3.5.1. The Principle of Stereographic Projection. ∗ represents stars in the celestial sphere, + represents stars projected on to the equatorial plane.

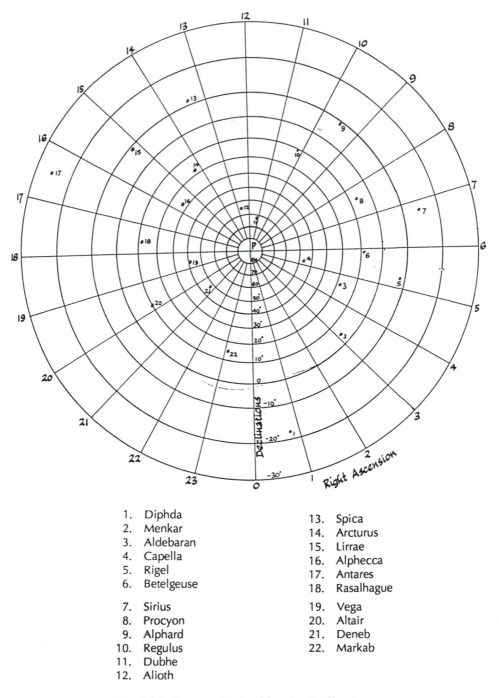

Fig. 3.5.2. Stereographic Star Map showing Key Stars.

1.	Diphda	13.	Spica
2.	Menkar	14.	Arcturus
3.	Aldebaran	15.	Lirrae
4.	Capella	16.	Alphecca
5.	Rigel	17.	Antares
6.	Betelgeuse	18.	Rasalhague
7.	Sirius	19.	Vega
8.	Procyon	20.	Altair
9.	Alphard	21.	Deneb
10.	Regulus	22.	Markab
11.	Dubhe		
12.	Alioth		

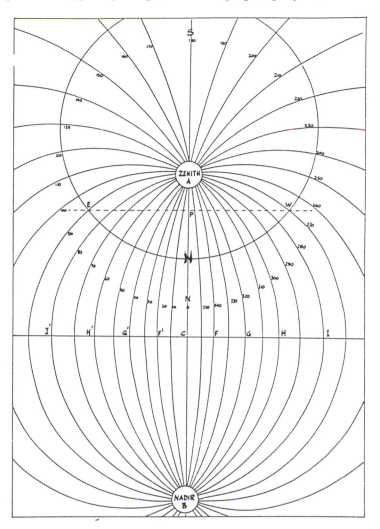

Fig. 3.5.3. Azimuth Lines drawn with a compass. When the full set of azimuth circles
has been drawn, a pleasing set of curves is produced. All the azimuth curves are parts
of circles. It is a matter of convention how the curves are marked. Usually they are
marked as shown above with zero Azimuth to the North.

The *Chart of the Heavens* published by the British Astronomical Association is
of this pattern, and can be constructed geometrically, using only a beam compass
and a ruler. The ancient Greeks had a profound interest in and respect for circles as
being *the perfect* form.

The British Astronomical Association's *Chart of the Heavens* is shown in Fig.
3.5.7. The *chart* is a stereographic projection — as used on mediaeval astrolabes
— all the curves are parts of circles, as shown in the upper part of Fig. 3.5.5. The
graticules superimposed on Fig. 3.5.7 are shown.

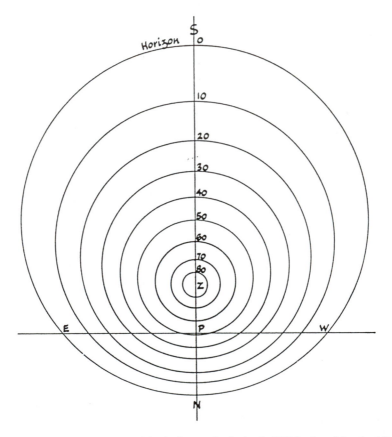

Fig. 3.5.4. Almucantars or Altitude Curves for Latitude 51° North and South. All altitude curves are circles but not concentric cirlces.

In order to represent stars below the celestial equator on this projection, an inconveniently large surface is required. So for stars of the middle heavens, having declinations between +40 and −40, a more convenient projection is used, namely the Mercator's projection, commonly used on geographical maps. A graticule appropriate for this section of the celestial globe is shown in the lower part of Fig. 3.5.7.

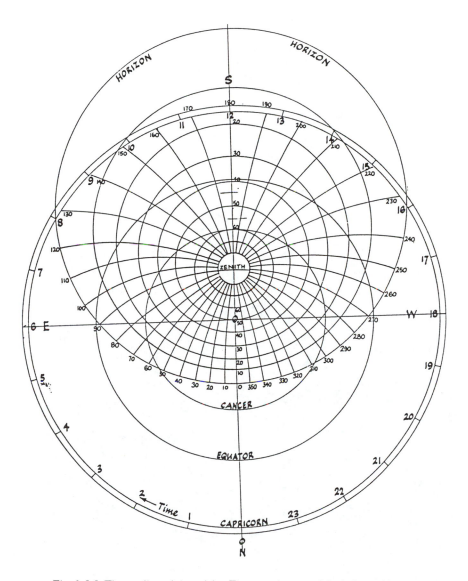

Fig. 3.5.5. The mediaeval Astrolabe. The complete astrolabe is formed by superimposing Fig. 3.5.3 upon Fig. 3.5.2 and both upon Fig. 3.5.4 giving this figure.

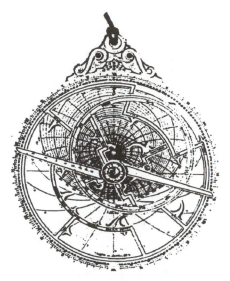

Fig. 3.5.6. The Medieval Astrolabe.

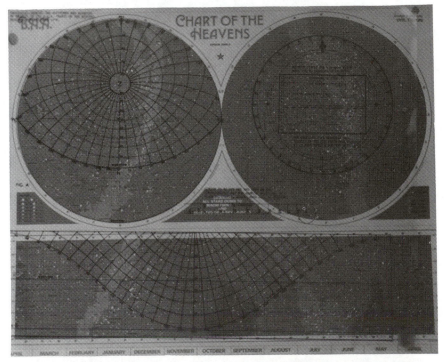

Fig. 3.5.7. The British Astronomical Association's Chart of the Heavens Superimposed on the Chart are two graticules formed by the author. Stars of the Northern celestial sphere having declinations from 40° to 90° are on a stereographic chart, and stars having declinations from +40° to −40° are on a Mercaror's chart. The graticules in each case show the altitutdes and Azimuths of the stars.

Although the two graticules appear to be quite different, they are each derived from the same alt–az formulae and the print-outs in Tables 3.3.1, 3.3.2 and 3.3.3, but used on different star map projections.

We have considered three kinds of planisphere, which can be used all the year round and can be set for any required time and date of the year. They cannot however show positions of planets, comets or the Moon. These positions, if required, may be marked temporarily in pencil, as they vary day by day.

3.6   STARS OF THE MONTH AND THEIR SPECIAL MAPS

Many skywatchers use the star maps such as those published at the beginning of each month in the *Daily Telegraph* or *The Times* or other newspapers, (Fig. 3.6.1).

These monthly maps with a little practice give a fair idea of the positions of celestial bodies shown on the map, but are suitable and accurate only at a particular time on a particular date — that is at a particular *Sidereal Time*.

This time, as we have seen in Section 1.12 is equal to the RA of the star that at that moment is on the meridian, i.e., when the LHA = 0.

For example, on the map shown in Fig. 3.6.1 for 23^h00 on December 1st, the Pleiades with RA $3^h 40^m$ is on the meridian, so the map is in position for Sidereal Time $3^h 40^m$. This can be checked using the nomogram shown in Fig. 1.13.4.

The interesting and particularly useful thing about these maps is that the RAs and declinations of its star positions are not plotted on the usual co-ordinate system of RAs and declinations, but instead on a system of altitude-azimuth co-ordinates depicting values that are applicable to the star positions only at a *specific* sidereal time. So in order to enhance the usefulness of these maps, all that is required is a small square of tracing paper, 140 mm side, a protractor and a pair of compasses to produce the alt–az overlay that can be used throughout the year, as shown in Fig. 3.6.2.

This graticule can be used to obtain the altitudes and azimuths of celestial bodies on the monthly star map at the beginning of each month. At the *end* of the month the graticule should be given a 30° rotation about its celestial pole — *not* about the Zenith of the map — in a clockwise direction. This rotation will represent approximately the position corresponding to the beginning of the next month.

For intermediate dates, use the fact that the graticule should be moved 1 for each day from the initial position on the 1st of the month.

The monthly star maps similar to Fig. 3.6.1 cannot easily be produced without considerable computational and graphical skill, but as published each month and used with the alt–az graticule Fig. 3.6.2 they give quick and useful results. Some may become a little confused by the fact that each monthly map is issued not only for 23^h00 each month, but is also applicable for six other times of the year. Although these times appear to be unrelated, they all in fact denote the same

Sidereal Time, which for more advanced sky watchers is useful and important, particularly for observing work with a telescope. It would enhance the usefulness of the maps if, additionally, a single Sidereal Time could be mentioned on each map.

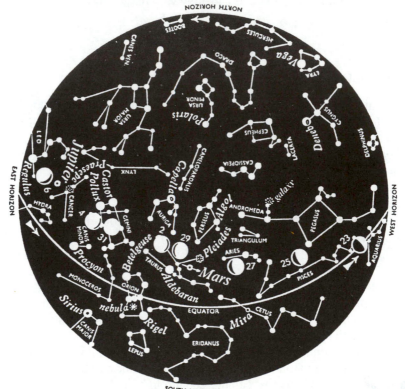

Fig. 3.6.1. Star Map for 23h00, 1st December, 1990 — Sidereal Time 03^{h}40. The aspect of the sky (apart from the Moon and Planets) will be approximately the same in other months at the following times — 05^{h}00, 1st September; 03^{h}00, 1st October; 21^{h}00, 1st January; 19^{h}00, 1st February — these times are expressed in Greenwich Mean Time, all these times and dates can correctly and succinctly be stated as Greenwich Sidereal Time 03^{h}40.

The Pleiades, RA approximately 03^{h}40^m is given by the graticule in Fig. 3.6.2, at this time on the NS meridian (Az 180°) and at an altitude of 60°.

Capella:	altitutde 75°	Azimuth 95°	(RA 05h16m Declination 45°59′)
Deneb:	altitude 27°	Azimuth 310°	(RA 20h41m Declination 45°15′)
Vega:	altitude 5°	Azimuth 330°	(RA 18h36m Declination 38°48′)

Verify this by using the overlay in Fig. 3.6.2.

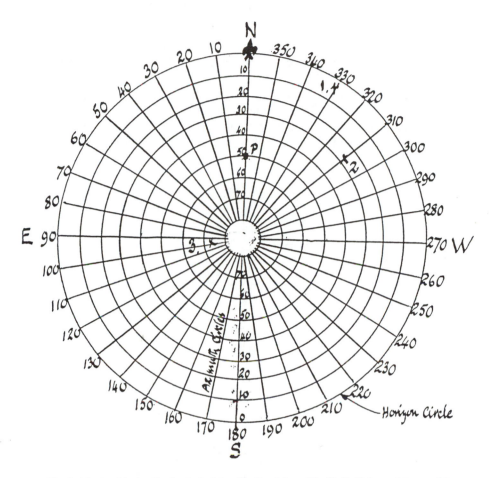

Fig. 3.6.2. An Alt-Az Graticule for Monthly Star Maps. The Daily Telegraph's monthly Star Maps by J.L.W. have been designed using Altitude-Azimuth co-ordinates instead of the usual RA-Declination co-ordinates. The maps can therefore be used to give, with accuracy, the altitudes and Azimuths of any of the celestial bodies shown on the map at the time specified. To do this, place the traslucent graticule exactly over the star map so that the horizon circles and the cardinal compass points coincide. Altitudes and Azimuths can then be read directly. When applied to the Star Map for 23^h00 on 1st December, the stars shown are:

1. Vega   Alt 5°   Az 330°
2. Deneb   Alt 27°   Az 310°
3. Capella   Alt 75°   Az 95°

A useful thing to do: Draw this overlay on tracing paper, and use on the maps given each month, to locate stars or planets in Altitude and Azimuth, on the hour and date specified, i.e., on the Sideral Time of the map 03^h40^m.

3.7   USING GRAPHS AND NOMOGRAMS TO SOLVE PROBLEMS OF STAR POSITIONS FOR SKY WATCHERS, AND FOR NAVIGATORS

Sky watchers often enquire, when looking at the stars, for example, ''What is the star or planet now in the sky bearing 69° at an altitude of 36° the time being 23^{h}00 LMT on 16th May?''

Stars are formally identified by their celestial 'addresses', in terms of their Right Ascension and Declination. There is no need to hold a star map or planisphere overhead to find these addresses as two graphs which can be constructed easily, or photo-copied, can be used to answer all such questions concerning star positions. Graphs that show relations between more than two variables are known as nomograms such as the ones shown in Figs 3.7.1 and 3.7.2. They can be used as follows:

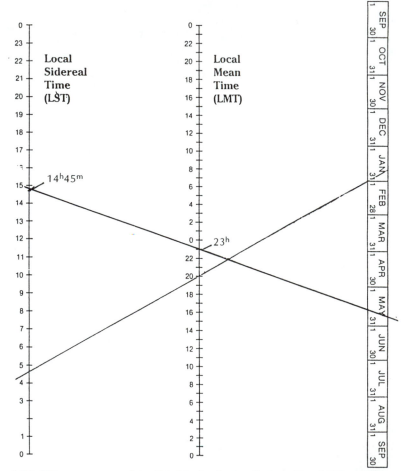

Fig. 3.7.1. This nomogram relates the date, local mean time, and local Sidereal Time. Any one of these values can be found when the other two are kown, by laying a straight edge across the three scales. The example shows that on January 31st, when the sidereal time is 4^{h}40^m, the local mean time is 20h (8:00 p.m.).

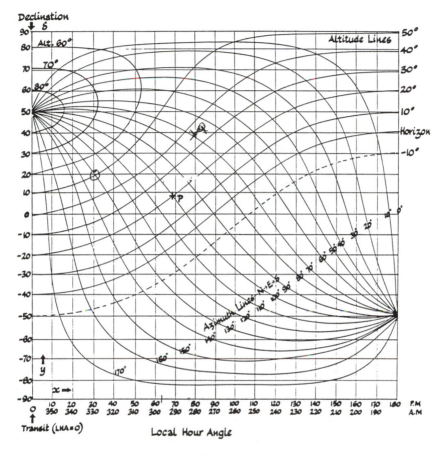

Fig. 3.7.2. The point Q marks the intersection of Altitude 36° and Azimuth 69° at Hour Angle
79° and Declination 40°.

Firstly, using Fig. 3.7.1 the date 16th May is aligned with the Local
Mean Time 23^h to give the Local Sidereal Time which in the example is
seen to be 14^h 45^m (ST is often required by observers).

Next, note the intersection of azimuth 69° and the observed altitude 36° marked
by the point Q, read the Local Hour Angle 79° = 5^h 16^m; the Declination is also
given at 40°.   The Local Hour Angle can be readily found from the relation

LHA = LST − RA

so the RA of the star is 14^h 45^m + 5^h 16 = 20^h 1.

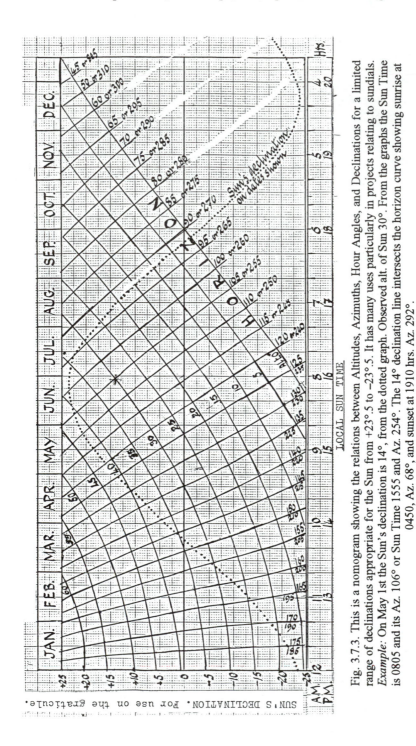

Fig. 3.7.3. This is a nomogram showing the relations between Altitudes, Azimuths, Hour Angles, and Declinations for a limited range of declinations appropriate for the Sun from +23°.5 to –23°.5. It has many uses particularly in projects relating to sundials. *Example:* On May 1st the Sun's declination is 14°, from the dotted graph. Observed alt. of Sun 30°. From the graphs the Sun Time is 0805 and its Az. 106° or Sun Time 1555 and Az. 254°. The 14° declination line intersects the horizon curve showing sunrise at 0450, Az. 68°, and sunset at 1910 hrs. Az. 292°.

The hour angle is measured westward along the celestial equator from the observer's meridian to the RA Hour Circle of a star or celestial body. It is expressed in hours and minutes from 0^h to 24^h; it is thus measured in the same units but in the opposite direction as Right Ascensions, so RA = $14^h 45^m + 5^h 16^m = 20^h 1^m$.

Another kind of question asked is similar to the one above, but in reverse. Where, using Earth co-ordinates altitude and azimuth, can a star RA 20^h and Declination $40°$ be found at 23^h00 on 16th May?

To solve this problem, firstly use the RA and Sidereal Time to obtain the Local Hour Angle of $79°$. This Hour Angle is then used in Fig. 3.7.2 with the Declination to fix the point Q which shows the Altitude to be $36°$ and the Azimuth to be $69°$. Armed with this information and using the simple measuring devices described in Chapter 2, this shows one exactly where in the night sky to find the star in question.

A convenient way to make use of the Sun to tell the time and to give compass bearings (Azimth) is to form nomogram s suitable for the Sun's limited range of dclinations, (see Fig. 3.7.3).

When used with care, these two Figs 3.7.1 and 3.7.2 provide answers to a wide range of problems and questions on star positions and astronavigation. The nomogram Fig. 3.7.2 encapsulates the computer printout Tables 3.3.1, 3.3.2 and 3.3.3 on a single sheet of A4 centimetric graph paper. The curves, too, are interesting to plot and the nomogram serves many uses for sky watchers, in finding, or identifying stars, planets or comets. Nomograms for Latitude $40°N$ and for Latitude $57°N$, are shown in Section 6.4, Figs 6.4.1 and 6.4.2.

3.8 STAR PATHS ACROSS THE SKY

Beginners in astronomy may have some difficulty in tracing the path of a particular star or planet as it apparently moves across the sky. Fig. 3.9.1 shows how the paths vary with the star's declination. With declination $0°$ a celestial body follows the line of the Equator across the sky from due East to due West. At declination $-20°$ (e.g., the Sun in Winter) a body describes a low arc, below the equator, and at $+20°$ declination it describes a high arc. The highest point reached on crossing the meridian is

$$(90° - \phi + \delta) \tag{1}$$

where ϕ is the latitude.

Stars having a declination $(90° - \phi)$ or more will not rise or set. They are called circumpolar, and are shown in the top part of the figure. There are two additional simple relations that are well worth checking using a calculator, which will enable the observer with equation (1) above to sketch with accuracy the azimuth (compass bearing) at rising or setting given by

$$\cos Az = \sin \delta / \cos \phi \tag{2}$$

Then the angle θ that a star makes with the horizon as it rises or sets given by

$$\cos \theta = \sin \phi / \cos \delta \tag{3}$$

Using the values obtained from equations (1), (2) and (3), the required sketches can be made, as shown in the lower part of the figure.

Four examples for latitude 51° are shown in the figure; they are for the Sun with declinations −23.4° (the Winter Solstice), 0° (the Equinox), +23.4° (the Summer Solstice) and a curve for a star of declination 39° which in latitude 51° just reaches the limit for being circumpolar. Curves for stars of other declinations can be sketched in by interpolating between the curves given and checked if necessary by calculation or by using a nomogram relating Altitude, Azimuth and declination.

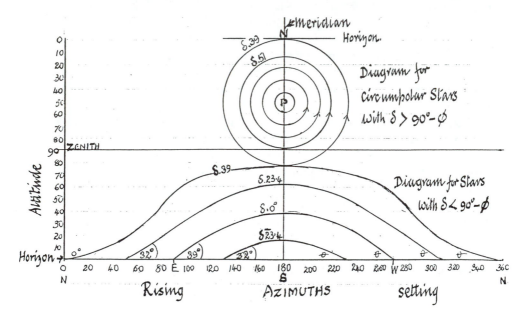

Fig. 3.9.1. Azimuths at rising or setting are given by

$$\cos Az = \frac{\sin \delta}{\cos \phi}.$$

The angle θ at rising or setting is give by

$$\cos \theta = \frac{\sin \phi}{\cos \delta}.$$

4

Light and Basic Optics

We receive our knowledge of the stars mainly through electromagnetic radiations including light, with which we are familiar, X-rays, ultra-violet rays, infra-red rays, and a wide range of radio waves.

Astronomy made great advances from about 1608 when the telescope was first used by Galileo, and still more spectacular advances from 1939 when radio telescopes were used. Sir William Bragg, over fifty years ago, said,

> *"Light brings us news of the Universe, coming from the Sun and the stars; it tells us of their existence, their positions, their movements, their constitution and many other matters of interest."*

4.1 REFLECTION

Here are some simple things to do concerning the reflection, refraction and scattering of light.

Hold a ball point pen **P** a few centimetres above a plate glass mirror **ABCD**, as shown in Fig. 4.1.1. **CD** is the silvered side. The eye at **E** will see a bright image of **P** by the ray **PORST** which changes its direction by refraction at **0**, is reflected at **R** and then changes direction again at **S**. The eye will see a very faint image at **P** from the ray **POG** which is reflected from **AB** at **O** the unsilvered surface, with the angle of incidence

$$POV = \text{the angle of reflection } VOG.$$

The faint image from **POG** is a ghost image close to the bright image, and for this reason it is quite unsuitable to use a piece of ordinary mirror for reflecting astronomical images in a telescope. Only if the surface **AB** is a perfectly plane surface and top silvered, can a good single image be formed when used to reflect an image in a telescope.

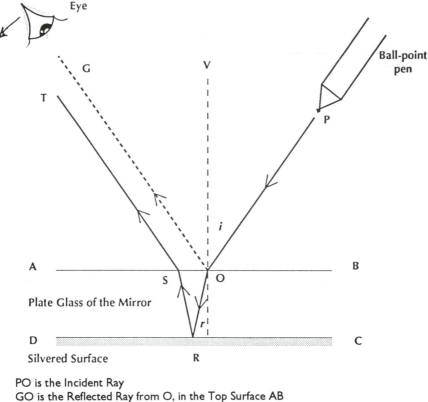

PO is the Incident Ray
GO is the Reflected Ray from O, in the Top Surface AB
TS is the Reflected Ray from the Silvered Surface CD,
    Refraction at O, Reflection at R and another Refraction at S

Fig. 4.1.1. Some things to do concerning the reflection and refraction of light.

4.2   REFRACTION OF LIGHT

This is a term used to describe the bending of a ray of light passing from one transparent medium to another; for example from air to water. A stick held partly immersed in water appears to suffer a bending at the surface separating the two media. A glass prism **ABC** shows refraction convincingly using pins to trace the path of the light (Fig. 4.2.1).

The prism **ABC** is placed on a bench with the plane **ABC** parallel to the plane of the bench. Now stick four pins P_1 P_2 P_3 and P_4 into the bench roughly in the positions shown. Look in the direction P_4 P_3 and adjust the position of the pins P_2 and P_1 so that the four pins P_1 P_2 P_3 and P_4 appear to be in a straight line. The actual path of the light ray P_1 P_2 P_3 and P_4 shows that there is refraction at P_2 air to glass, and another refraction at P_3 glass to air. When P_2 P_3 is parallel to the base of the regular prism the deviation (i.e., the angle d between **P₁ P₂** produced (**P₂ P₅**) and **P₃ P₄** produced) is a minimum.

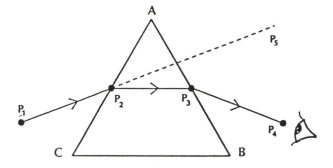

Fig. 4.2.1. A glass prism showing refraction.

It will be noticed that the pins **P₁ P₂ P₃** will be slightly coloured along the edges, this demonstrates another important effect. Not only does the prism refract the light from the pins, but it also bends red light slightly less than blue light. This effect is of great importance in optical instruments and was first demonstrated by Newton. He arranged for a narrow beam of light **AB** (a narrow slit of sunlight) to enter a prism as in Fig. 4.2.2. This narrow beam is split on refraction at **B** and further split at **C**, into a coloured beam ranging from red at **D$_R$** to violet at **D$_V$**.

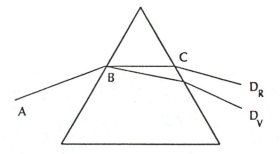

Fig. 4.2.2. The refraction of a narrow beam of sunlight.

The Sun may not be easy to use for this demonstration but a bright slit on a slide can conveniently be used with the aid of a slide projector. The "slit" slide can be made from two halves of an over-exposed film slide separated by a gap of about 0.5 mm as shown in Fig. 4.2.3. Focus the slit on a white screen about 3 m away and then place a prism in front of the projector lens to produce a strikingly beautiful spectrum, 1.3 m to one side of the direct white image of the slit. The angle of deviation

$$= \text{Arc} \tan \frac{1.3}{3} = 23°.4$$

A suitable prism for this demonstration can be made cheaply by joining together 4 pieces of thin glass or perspex each 75 mm × 75 mm × 2 mm as shown in Fig. 4.2.5 using a liberal quantity of silicone rubber adhesive, so that the prism is water-tight. The model is then filled with water. The angle of incidence for a 'water' prism refractive index (yellow light = 1.34) is 42° and the angle of minimum deviation = 24°.

Fig. 4.2.3. How to make the narrow slit.

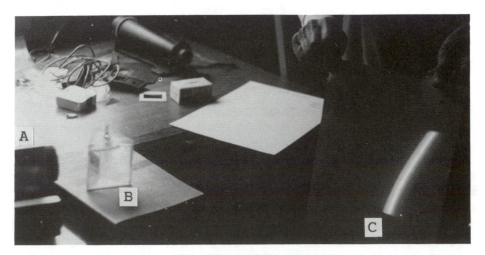

Fig. 4.2.4. Showing the spectrum at C of Light from a slide projector lamp at A, passing through a slit (Fig. 4.2.3) and a prism B.

Fig. 4.2.5. The construction of the prisms.

Having produced a brightly coloured spectrum by splitting or dispersing white light into violet/indigo/ blue/green/yellow/orange/red, the seven recognised colours, Newton had the bright idea of placing a second prism in the path of the dispersed light as shown in Fig. 4.2.6.

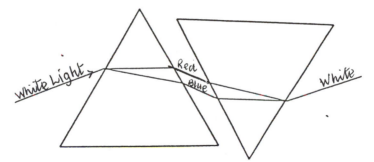

Fig. 4.2.6. The recombination of white light.

The second prism has the effect of recombining the colours so that white light emerges; a convincing demonstration that light can be split into its component colours, but can also be made to recombine to form white light. This fact is made use of in making camera and telescope lenses that are practically free from unwanted colour fringes, using a combination of lenses and two different kinds of glass selected and shaped with great skill and precision. The result is called an **achromatic combination** and such combinations in lenses make them very expensive.

The way in which a prism refracts light depends on the angle of the prism and explains how a lens brings a parallel beam of light to a focus after leaving the lens.

We can regard the lens as being made up of a large number of prisms **ABCDE** as shown in Fig. 4.2.7. At **A** the angle is greater than at **B**. But the ray **OAI**$_Y$

reaches I_Y by way of **OB**I_Y. The part of the lens at **C** has parallel "sides" so **OCI** is a straight line (The Optic Axis). All rays from **O** received by the lens meet at I_Y. I_Y is the focal point for yellow light.

We have used the prism to show that red light is refracted a little less than violet light. This means that red light is brought to a focus at I_R, a little further away from the lens than I_Y (yellow) while violet light is brought to a focus at I_V which is a little nearer to the lens than I_Y. The achromatic lens that has been described, corrects this dispersion of colours.

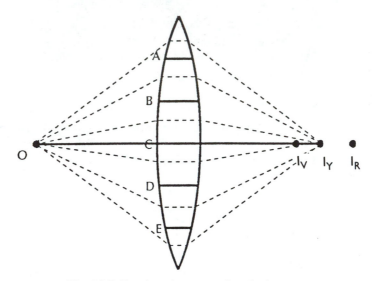

Fig. 4.2.7. Treating a lens as a series of prisms.

4.3   TELESCOPES

For sky watchers today, as in the days of Galileo, passing from naked eye to telescope viewing is an exciting experience. The naked eye can perceive about 1000 stars, most of them just bright enough to be seen, but with a small telescope or binoculars hundreds of thousands of stars, as well as nebulae, galaxies and clusters, become surprisingly visible. The satisfying thing about this, is to understand how a telescope works, and find out what it can do for us, and how it can show stars too faint for us to see with unaided eyes. Many false notions are held and written about telescopes especially in advertisements which claim very high magnifications that are either false, or the images they produce are so poor or faint as to be of no use.

In this section we examine what a telescope can do under good conditions. See Table 4.19.

4.3.1   Light gathering power of a telescope

First we deal with the light gathering power of a telescope.

The most important function of a telescope is to gather light by means of its object lens or mirror from a distant object such as a star, or planet. The amount of light it can gather is proportional to the area of the object lens or mirror, i.e., to D^2 where D is the diameter of the object lens or mirror. A good instrument brings this light from a star to a focus. This is then looked at through an eyepiece which is really a simple microscope, such as a watch-maker's eye glass, which is an item of study in physics syllabuses. If you wish to enquire about the capability of a telescope. don't show beginner's ignorance by asking "What is the magnification?" But ask. "What is the diameter in millimetres of the object lens?" (see Section 4.3.2). The pupil of your eye has a maximum diameter, when dark-adapted, of about 8 mm. We use this fact to find the *light gathering power* of a telescope, which has an aperture of D mm diameter.

Star brightness magnitudes are ranked rather like a football league system. The bright stars are in the 1st Division or in the magnitude 1 class, less bright stars are in the 4th Division or 4th magnitude. This star classification is a very old one and has a mathematical basis but it is a little confusing for beginners who may feel that a star magnitude 6 should be brighter than a star magnitude 1.

The light gathering power of a telescope is expressed as the ratio

$$\frac{\text{Area of telescope aperture}}{\text{Area of eye aperture}} = \frac{D^2}{8^2} .$$

For a modest telescope having $D = 75$ mm, the light gathering power is thus

$$\frac{75^2}{8^2} = 87.89$$

i.e., it collects about 88 times more light than can enter the unaided eye. We shall see later that a star barely visible without the telescope, and so a poor performer in the star brightness league (Division 6), is promoted by the light gathering power of the telescope to the top Division 1 (Magnitude 1) that is, it appears as bright as Vega. The Table under Light Gathering Power has a simple formula

$$M = 2 + 5 \log_{10} D$$

which is derived in Section 5.9. This formula gives the magnitude M that is the limit for a particular telescope. A telescope of 100 mm diameter objective will reveal stars that have a magnitude of 12, using this formula, we have $(\log_{10} 100 = 2)$

$$M = 2 + 5 \log_{10} 100 \text{ which in turn equals } 2 + 5 \times 2 = 12$$

In Section 5, Fig. 5.9 shows a graph illustrating the relation between M and diameter D of the objective lenses of telescopes.

4.3.2   A simple refracting telescope used on the full Moon

A simple diagram (Fig. 4.3.2) will illustrate how a small refractor telescope works when focussed on the full Moon which is a convenient object to experiment with when using a telescope. It provides plenty of light and the Moon has an angular diameter of about 0.5°.

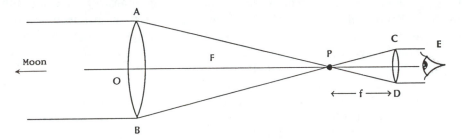

Fig. 4.3.2. How a small refractor telescope works when focussed on the full Moon.

An image of the Moon, **d** mm in diameter, is formed at **P**, the principal focus of the object lens **AB**, of focal length **OP** = **F**. The angular diameter is determined by the equation,

$$\frac{d}{F} \tag{1}$$

CD is the eyepiece of focal length **f**, which acts as a simple microscope and enables the eye, placed close to the eyepiece to see the Moon's primary image at **P** with an approximate angular diameter of

$$\frac{d}{f} \tag{2}$$

This is greater than (1) by a factor of

$$\frac{F}{f}$$

and this is the magnification of the telescope. By similar triangles **ABP** and **CDP**,

$$\frac{AB}{CD} = \frac{F}{f}$$

So the magnification of a telescope can be measured by the ratio

$$\frac{\text{Diameter of object lens (or mirror)}}{\text{Diameter of exit (beam)}}$$

This can be verified for any telescope or binoculars by pointing the instrument towards the sky or a bright surface and measuring the diameter of the exit beam using a millimetre scale. If this, for example, should be 2 mm and the diameter of the object lens aperture is found to be 150 mm. Then the magnification is

$$\frac{150}{2} = 75$$

which is a useful magnification for this instrument, and the focal length, **f**, of the eyepiece, for a telescope of focal length of 1000 mm would be

$$f = \frac{1000}{75} = 13.3 \text{ mm}$$

Cheap instruments claiming fantastic magnifications should be rejected if they fail these simple tests.

4.3.3   Magnification and the accommodation of the eye

We have seen that the telescopic magnification of a distant object such as the Moon or a planet really consists of two quite separate parts as described in Section 4.3.2.

(1)   The object lens or mirror produces, by itself, a primary image which can be formed on a screen, or on the film of a single lens reflex camera, which can be looked at directly by the unaided eye. An object lens of, say, 1000 mm focal length produces an image on a screen which actually has magnification

$$Mo = \frac{F}{250} \quad \frac{1000}{250} \text{ times}$$

So Mo = 4 as seen by the eye without the eyepiece. This is the size that the image appears to be to the unaided normal eye, since the eye sees things best at a distance of clear vision which is normally 250 mm. So an object lens of focal length 1000 produces the *real image* which is magnified four times.

(2)   The eyepiece of focal length f now takes over and acts as a simple microscope, and forms a *virtual image* of the primary image at a distance of distinct vision, normally 250 mm, from the eyepiece and so magnifies the primary image acting as a simple microscope

$$M_e = \left(\frac{250}{f} + 1 \right)$$

as given in textbooks for a simple microscope derived from the formula for a lens

$$\frac{1}{v} - \frac{1}{u} = \frac{1}{f}$$

where **u** = the distance of the object from the lens, and **v** = distance of the image, which gives the magnification which can be expressed as

$$\frac{u}{v}$$

The Total Magnification of a telescope is thus $M_o \times M_e$

$$= \frac{F}{250}\left(\frac{250}{f} + 1\right) = \frac{F}{f} + \frac{F}{250}$$

for a normal eye. This approaches

$$\frac{F}{f}$$

which is generally accepted since

$$\frac{F}{250}$$

is small compared with

$$\frac{F}{f}$$

and also the distance of clear vision depends partly on the accommodation of the observer's eye. Many works on telescopes ignore this small contribution to the magnification.

Telescope users are often bothered when wishing to demonstrate their telescope's performance, to find that what is a sharp focus for one person is out of focus for another. This is on account of the F/250 factor and is corrected by racking the eyepiece in a little for short-sighted friends or out for long-sighted ones. Only a millimetre or so is generally required for this adjustment.

4.3.5   Telescope mountings

Telescopes are designed and mounted to be able to point to any part of our visible celestial hemisphere which extends in two directions that are at right angles to each other and with which we Earthlings are very familiar, namely the vertical (plumbline) and the horizontal (horizon). Telescopes are mounted using two axes that are at right angles to each other.

For example, most small telescopes are mounted on a main vertical axis **AB** which permits the telescope to point in any direction of the compass, known as bearing or azimuth. Combined with this axis is another axis **CD** pointing in any altitude from level (the horizon) to overhead, (the zenith). This mounting is commonly called an altazimuth mounting.

Fig. 4.3.5 shows a D.I.Y. altazimuth mounting for a fairly heavy telescope. The mounting is known as **Dobsonian**.

The two pillars shown are firmly fixed to a disc which is free to turn about a vertical axis, resting on ball bearings. The ball race can be formed from two cycle wheel rims that are slightly different in diameter screwed on to a base plate **P**. The space is filled with glass marbles, and the disc with pillars rests centrally on the ball bearings. The pillars support a cradle for the telescope — any telescope — and the cradle can turn on two trunions, about a horizontal mounting lined with polythene which has a very low coefficient of friction. The bulky telescope shown can be turned by finger pressure to put it in any azimuth and at any altitude and it will remain steady and stable. We unconsciously adopt an alt-az mounting when using binoculars: as we can turn ourselves about a vertical axis and use the hinge in our neck to look from the horizon to overhead, rather uncomfortably. To avoid discomfort when using binoculars, or any alt-az instrument, see Section 4.15.

When using binoculars, it is well worthwhile to do so with an accurate knowledge of the location of the North, East, South and West points on your horizon. This enables you to make fair estimations of the azimuth of objects you are observing. Altitudes can readily be measured using devices such as Fig. 1.9.1.

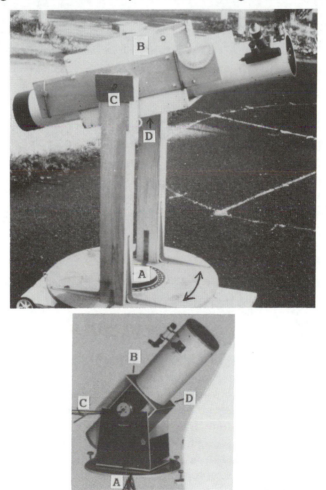

Fig. 4.3.5. A Dobsonian (alt-az) mounting for a fairly heavy telescope.

Fig. 4.3.6. Marbles on a ball-bearing race giving an efficient azimuth turning track.

4.4   THE EQUATORIAL TELESCOPE

This is similar to the alt-az instrument in that it has two axes fixed at right angles to each other. If the principal axis of the alt-az telescope (normally vertical), **AB** is inclined to the horizontal at an angle equal to the latitude in the plane of the meridian, then the azimuth circle becomes the ''Right Ascension'' circle, which can measure the angles that stars make with the plane of the equator as shown in Fig. 4.4.1, and the altitude circle then becomes the declination circle of an equatorial telescope.

4.5   REFRACTORS AND REFLECTORS

Telescopes are optically of two kinds: refractors and reflectors. The simple model described in Fig. 4.5.1 is called a refractor telescope because the primary focus is formed by a lens which, as we have also seen, refracts the incident light to form an image (Figs 4.2.7 and 4.3.2). We` have also seen that refraction always suffers some degree of colouration or dispersion of light first studied by Newton (Fig. 4.5.1). He overcame this colour problem at the primary stages by bringing a light beam to a focus not by refraction through a lens **AB** as in Fig. 4.5.1, but by a concave mirror, **BC** a *reflector* (see Fig. 4.5.2) which, if properly figured brings a beam of parallel light, e.g., that from a star, to a point focus **I**. The shape for this is a parabolic mirror which acts on a parallel beam as in the figure 4.5.2 (also see Section 5.12 on grinding your own mirrors where the principles are dealt with in detail).

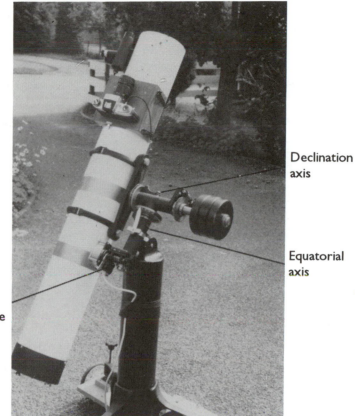

Fig. 4.4.1. An equatorial reflector telescope, showing its two axes.

The optics for refractors and reflectors are similar in that they each provide a principal focus **I** as shown in both Fig. 4.5.1 and Fig. 4.5.2. This image is observed in a refractor telescope by means of an eyepiece directly, but in the case of the reflector telescope, the image of **I** has to be deflected by a small plane mirror **M** and is observed through an eyepiece at **E** from the side of the telescope tube. The refractor has an objective lens of focal length **F**, while the reflector instrument has a parabolic mirror of focal length **F**. We are familiar with the properties of a parabolic mirror in reverse as car headlights have a light bulb at the focus of a parabolic mirror which projects a parallel beam of light (see Section 5.12).

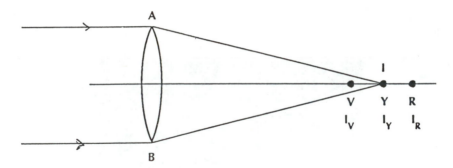

Fig. 4.5.1. A refractor showing a degree of colouration, or dispersion, of light first studied by Newton.

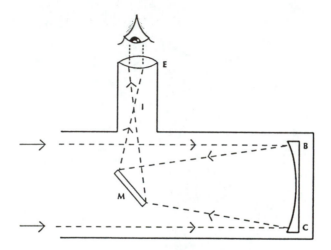

Fig. 4.5.2. A reflector, showing its parabolic mirror BC.

4.5.3 Measuring altitudes and azimuths with an alt-az telescope

Many small alt-az telescopes, although they are mounted so that they move about a horizontal axis for altitude and a vertical axis for azimuth, rarely seem to have the means of measuring these important parameters, possibly because transforming altitudes and azimuths into Right Ascensions, Declinations and Hour Angles involves the local time and Sidereal Time and resort to spherical trigonometry and tables. With the use of calculators, computers or planispheres with graticules these transformations can readily be made.

An alt-az instrument can be used to measure azimuths by mounting it at the centre of an azimuth table which is a circular disc (see Fig. 4.5.3) marked in degrees round the periphery, from 0° at North, to 90° at East, 180° at South and 270° at

West. This is a very useful disc, 50 cm in diameter, as it can serve as a platform for various kinds of small instruments that are concerned with azimuths. It should have a central hole that can take a 10 mm bolt and nut to form a vertical axis. The disc can be supported by three legs as shown, which can facilitate levelling when necessary. The simplest and most effective means of measuring or setting a tele-scope for altitude is a semi-circular protractor mounted on a sighting arm which is fastened on the telescope parallel to its line of sight (see Fig. 4.5.3). A fine thread passing through a hole at the centre of the protractor, and carrying a small weight, serves as a plumb line which indicates the altitude of the telescope to within half a degree.

Fig. 4.5.3. An alt-Az reflector telescope mounted on an azimuth table in compass bearing position with a protractor and plumb-line for altitude observations.

The photograph in Fig. 4.5.3 shows the alt-az accessories on a telescope. A sighting bar and protractor can be used to find the altitude of any object by holding it in the hand. For finding the altitude of the Sun, the shadows of two small projections on the sighting bar are brought into line **to avoid looking directly at the Sun**.

4.6   USING TELESCOPES

There are many elaborate instructions and methods used for setting up an equatorial telescope in your observatory, or backyard, but two basic requirements are essential.

(1)   The telescope must be securely mounted so that its equatorial axis (i.e., the polar axis) is accurately in the N–S meridian and pointing to the celestial pole (altitude = Latitude of place).

(2)   The declination axis must be at right angles to the polar axis. This means that when the telescope points due South, the declination axis is horizontal. To check the declination circle, point the telescope due South towards the horizon, so that the telescope is horizontal as checked by a good spirit level. The declination circle should then read minus $(90° - \phi)$ where ϕ is your latitude.

There are many more things to do at the telescope such as checking positions of stars (the Pole star is a good test) or checking local Sidereal Time by observing stars of known RA in transit or by applying the nomogram shown in Fig. 3.7.2. Proper use of setting circles can provide much pleasure and satisfaction, in conjunction with charts and planispheres, when looking for, or identifying, celestial bodies.

4.6.1

A handy D.I.Y. set square for telescope users and sundialists is shown in Fig. 4.6.1. **ABC** is a right-angled triangle of 3 mm perspex conveniently cut from a rectangle of the material. One side of the triangle is 250 mm the other is of length L such that

$$\frac{250}{L} = \tan \phi$$

where ϕ is the latitude of your backyard, or observatory. My backyard is latitude 51.1° N so in the figure

$$BC = \tan \frac{250}{\tan 51.1°} = 207.7 \text{ mm}$$

Perspex is easy and satisfying to cut with accuracy using a hacksaw, and costs only a few pence. As the figure will show from its markings it can be used to set or to check various angles on and around the telescope.

A plumb-line hangs perpendicularly from **A** on to the horizontal **BC** which when aligned in the meridian ensures that the line **BE** is in the plane of the ecliptic, and the angle **CBE** is the altitude of the equator above the horizon at the South point. A telescope pointing in this direction with its declination circle properly set will read Declination 0. A small spirit level is fixed with a little adhesive to level **BC** when wind upsets the plumb-line. An additional embellishment is a small circular plastic protractor for use in checking azimuth observations.

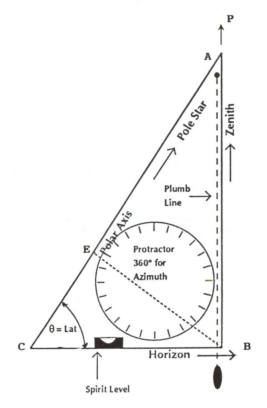

Fig. 4.6.1. A D.I.Y. set square for telescope users.

4.7   SETTING CIRCLES

There is a tendency among inexperienced observers who own or use telescopes to regard the Right Ascension and declination circles as being interesting pieces of ornamentation of little or no functional value. Such users rely entirely on "star hopping", in order to aim their telescopes to a desired RA and declination. Neglect of setting circles can deprive observers of expert experience and understanding of star positions and movements and how they are related to star charts, planispheres and star globes.

In the Fig. 4.7.1 consider, first of all, the RA setting circle **B** which encircles the polar axis. **A** is a pointer which is fixed to the casing of the telescope's equatorial axis **PC** and is thus in the plane of the meridian. At **B** is a small pointer which is fixed to the moving part of the telescope that turns about the polar axis. This pointer is set on the RA of the body to be observed. This, in the photograph, is seen to be 16^h45^m.

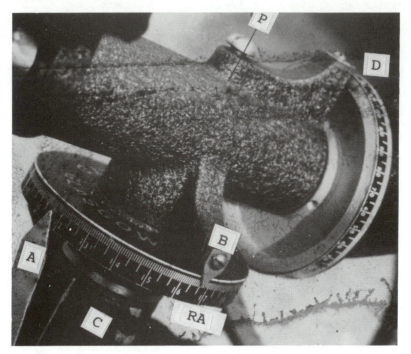

Fig. 4.7.1. Setting Circles. The photograph shows the RA circle on the main axis. The point A is on Sidereal Time and is fixed due South of the instrument. B is on the star's RA. Clearly LHA = ST − RA.

The next step is to turn the telescope about its polar axis until the pointer **A** reads the Local Sidereal Time (LST). 11^h45^m. This is Greenwich Sidereal Time − Longitude West or + Longitude East.

The telescope is now set for the RA of the body. For the declination, turn the telescope about its declination axis so that the pointer **D** reads the declination of the body. As in section 3.2, the Right Ascension of any star on the meridian is shown by the reading of pointer **A**. If the telescope is clock driven to keep sidereal time, it registers the RA of stars as they transit the meridian, and the telescope will keep the selected star in its field of view for as long as required. A clock drive is essential for astrophotography that may require an exposure of more than a few seconds.

The mounting of an equatorial telescope with its setting circles can be demonstrated by means of a simple model shown in Fig. 4.7.2. This shows the "telescope" **AB** which can turn about the axis **CD** which points to the celestial pole, and is therefore parallel to the Earth's axis. The pointer RA is in the plane of the telescope and can move over the white disc which is marked with the 24 hours of Sidereal Time round its periphery. The telescope can turn about an axis that is perpendicular to the polar axis, and can move over a protractor marked to indicate declinations on the celestial sphere.

The figure 4.7.2 shows the Right Ascension setting circle set for Sidereal Time 22^h30^m which corresponds to the Local Mean Time of 20h on 31st October. The telescope is set for RA 19^h50^m and for declination $8°\ 50'$ and so is pointing to the star Altair following the procedures outlined in Section 4.7 for setting a telescope and using setting circles.

Fig. 4.7.2. A model to demonstrate the mounting of an equatorial telescope.

4.8   THE MOON

The Moon provides us with several rewarding things to do using a small telescope. At the full Moon its craters and mountains are well illuminated for observing or photographing, and images provide a very convenient size for studying the capabilities of your telescope (its field of view and its magnification **M**).

Our Moon appears to us to have a constant angular diameter of approximately $0.5°$. Incidentally, this size is practically the same as that presented by the Sun's disc and so accounts for the total eclipses of the Sun by the Moon. If the Moon, for example, just fills our field of view at the eyepiece (which has a field of view of say 40°, see Section 4.12) then the field of view is 0.5°, so that, 40/M = 0.5° and M = 80°.

To use the Moon to determine the magnification of your telescope objective which, say, is of focal length F = 1000 mm, remove the eyepiece of your telescope and point the telescope at the Moon. Hold a piece of tracing paper at the primary focus of the objective lens, where a clear image of the Moon will appear. As the Moon subtends an angle of 0.5° at the lens, the image will subtend an angle of 0.5° at the centre of the lens 1000 mm away as shown in Fig. 4.8.1.

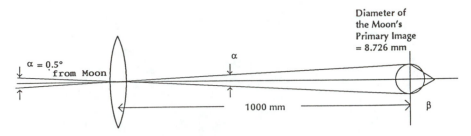

Fig. 4.8.1. Using the Moon to determine the magnification of your telescope objective.

From this Figure it can be seen that

$$\frac{d}{1000} = \frac{0.5°}{57.3} \text{ Radians}$$

so d = 8.726 mm which is the diameter of the primary image of the Moon.

Now draw a circle having this diameter and hold it at a distance of 250 mm from your eye but in the line of sight of the Moon, so that you can compare the size of the drawn circular primary image at 250 mm with the real Moon's disc. and so confirms the optics given in Section 4.3.3 and shows that the object lens *alone* produces a magnification of about 4, given by

$$\frac{F}{250} = \frac{1000}{250} = 4$$

as given in Section 4.3.3 Equation (1).

4.9   LENS FORMULA FOR EYEPIECE PROJECTION

The arrangement shown in Fig. 4.9.1 will be recognised by photographers as very convenient for taking enlarged photographs of the Moon. This is done by attaching a camera, minus its lens, to the eyepiece **E** which then serves as the camera lens. It will be noticed that the distances **u** and **v** can be varied to provide a wide range of magnifications on the film at **PQ**. For example, we have seen that when **u** = 25 mm it makes the image at the camera film equal to 35 mm — too large to be received on a 25 mm × 35 mm film. Whereas with **u** = 27 mm **v** will be 78 mm and the magnification will be

$$\frac{78}{27} = 2.86$$

giving the image of the Moon a size of 2.86 mm × 8.726 mm = 25 mm in diameter.

  This is a convenient size of image of the Moon to fill a 25 mm × 35 mm film.

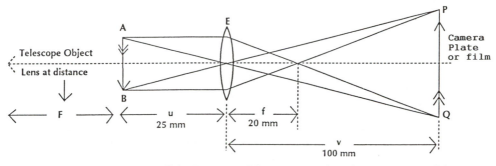

AB is the Primary Image, E the Eyepiece is 25 mm away, and produces an image at PQ.

Fig. 4.9.1. Eyepiece projection and photography.

  The Moon is thus a useful object on which to practice astrophotography. For those who like figures, the relation used in these lens calculations is

$$\frac{1}{v} + \frac{1}{u} = \frac{1}{f}$$

so that with **f** = 20 mm, **u** must be a little more than **f**. Then **v** can be calculated and

$$\frac{v}{u}$$

gives the magnification of the primary image **AB** at the camera film. If we make **u** = **f** then **v** and the image becomes infinitely large but too faint to be seen.

  In Section 4.10 the magnification for photography is conveniently given by

$$\left(\frac{v}{f} - 1 \right)$$

The magnification can be considered as equivalent to making the telescope longer, so we say that the effectual focal length **F** becomes

$$F \left(\frac{v}{f} - 1 \right)$$

and for photographers the effective focal ratio or its ''f number'' for the telescope objective lens diameter **D** is

$$\frac{F}{D}\left(\frac{v}{f} - 1\right)$$

4.10   TAKING PHOTOGRAPHS WITH YOUR TELESCOPE

From the dimensions given, the diameter of the primary image as seem at 250 mm from the eye is about four times the diameter of the Moon 0.5° observed directly by eye. The angle that 8.726 mm makes at a distance of 250 mm is

$$\frac{8.726}{250}\text{ radians}$$

(one radian = 57.3°) so the angle that 8.726 mm makes with the eye at 250 mm is equal to

$$\frac{8.726}{250} \times 57.3 = 2° \ .$$

Now we use the eyepiece, which is really a small microscope, to look at the primary image. The eyepiece has to be racked in or out until it is at a distance **f** from the primary focus where **f** is the focal length of the eyepiece. In this position it produces a virtual image (virtual because the eye can see it, but it cannot be caught on a screen) which can be seen when the normal eye is relaxed and focussed for looking at very distant objects. By racking the eyepiece inward a millimetre or so, a clear image will be seen at a place favoured by normal eyes known as the **normal distance of distinct vision**, 250 mm from the eye.

While experimenting with positions of the eyepiece (pushing it in for your short-sighted friend and out for your long-sighted uncle) try racking the eyepiece out 5 mm so that the eyepiece **E** produces an enlarged image of the primary image on a small screen placed at a point 100 mm from the eyepiece, as in Fig. 4.9.1 The primary image of the Moon **AB**, diameter 8.726 mm will be enlarged to **PQ** on the screen by eyepiece projection by a factor of 4.

4.11   MAGNIFICATION BY EYEPIECE PROJECTION

It is of interest to note that magnification of the primary image **AB** by eyepiece projection is

$$\left(\frac{v}{f} - 1\right)$$

so when:

v = 2f	**M = 1**
v = 3f	**M = 2**
v = 4f	**M = 3**
v = 5f	**M = 4**

These results can be verified satisfactorily on the Moon and planets but **great care must be exercised in forming images of the Sun** (see Section 2.3). The primary image of the Sun formed by an object lens of $F = 1000$ mm is, as we have seen, the same diameter as the Moon, namely 8.7 mm. **This is a very hot spot, so keep eyes and/or films well away from it!** The thing to do here is to project this small image on to a screen in a darkened box at such a distance that the size of the Sun's image is about 150 mm. This is a convenient standard size used in the study of Sunspots. Fig. 4.9.1 can be applied to the Sun but **PQ**, the projected image on a screen, has to be enlarged so that **PQ** = 150 mm. This is achieved by fixing the screen at a distance **v** from the eyepiece **E** so that the magnification is

$$\frac{150}{8.7} = 17.24$$

so

$$\left(\frac{v}{f} - 1\right) = 17.24$$

where $f = 20$ mm. Therefore $v = (17.24 + 1)20 = 36.48$ cm. This fixes the length of the projection box.

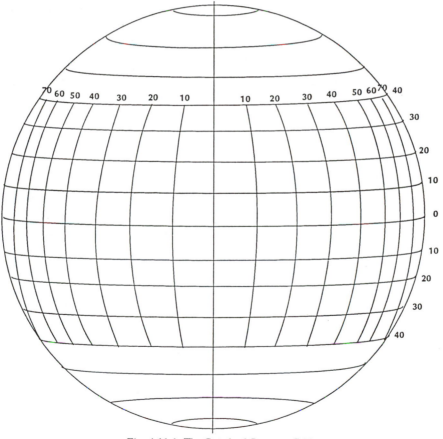

Fig. 4.11.1. The Standard Sunspot Grid.

The projected image can be allowed to fall on a standard Sunspot grid shown in Fig. 4.11.1 marked in co-ordinates that help to identify the positions of Sunspots in terms of the Sun's latitude and longitude. Images of the Sun by projection are quite safe to carry out, and they can be photographed by a camera suitably mounted, so that it is pointing to the image as nearly as possible perpendicularly to the plane of the screen, (see Section 2.3.2).

4.12   FIELD OF VIEW IN AN EYEPIECE

Look with one eye at the night sky and, keeping the head and eye still, try to estimate the angle of your field of view. A rough estimate for most people is about 90° to include the topmost and the lowest star that just comes into the view. When you use an eyepiece the field of view is generally about 40°. This can be tested as follows, as shown in Fig. 4.12.1.

ABF is a protractor to which is attached by means of a little 'blutac', midway between **A** and **B**, the eyepiece **E**. Hold the protractor horizontally and, with one eye, look at the eyepiece **E** and note its circular field of view. Rotate the protractor so that the eye is at **C**, which is the point at which the field of view becomes extinguished, and note the point **C′**. Finally, turn the protractor anti-clockwise until the field of view of the eyepiece is again extinguished at **D**. Mark the point **D′**. Then **C′ED′** is the angle of the field of view and for many eyepieces this is about 40°.

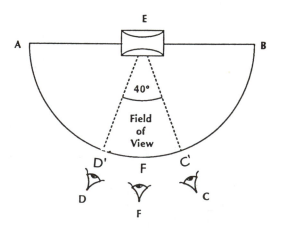

Fig. 4.12.1. Testing the field of view of an eyepiece.

4.13 BINOCULARS

Binoculars are specially useful and a great convenience for sky watchers. They are generally marked with two numbers such as 10×50 which indicate that the magnification is 10 and the diameter of the objective lens is 50 mm. The exit beam is accordingly

$$\frac{50}{10} = 5 \text{ mm.}$$

which can be effectively used by the eye pupil which has a full diameter of about 8 mm. For as was stated in Section 4.3.2,

$$\frac{\text{Diameter of the Objective Lens}}{\text{Diameter of exit pupil}} = M.$$

The image of the Moon in any telescope has a useful property in that it occupies a field of view of 0.5 of a degree. The greater the magnification, the less area of the sky will appear in the field of view of the telescope. If, for example, you attempt to look at the pointers of the Plough, Merak and Dubhe with a pair of binoculars of magnification 10, you will not be able to see both stars together as these two stars are separated by an angle of $5° 22'$ whereas the binoculars have a field of view of the sky

$$\frac{\text{Eyepiece field}}{\text{Magnification}} = \frac{40}{10} = 4°.$$

To see both stars at the same time you would require binoculars of magnification $8\times$ or less. A simple means of finding or checking the field of view of an eyepiece is described in Section 4.

To find the field of view of your telescope using a stop watch, we use the fact that stars appear to move across the sky near the celestial equator at the rate of $1°$ every 4 minutes approximately. So if a star near the equator moves across the telescope field of view in, say, 2 minutes, then the field of view is 0.5 of a degree, and assuming the field of the eyepiece is $40°$ then the magnification, **M**, is

$$M = \frac{40°}{0.5°} = 80.$$

Stars that are not near the equator, having a declination will appear to move more slowly, since they move in smaller circles of declination, so the time for a star of declination δ is given in minutes by the calculation

$$\frac{40 \times 4}{\cos \delta}.$$

At the celestial pole, $\delta = 90°$ and $\cos \delta = 0°$, so the time to traverse the field of view is infinite. So, as we would expect, the star will never make the crossing. (See Fig. 3.9.1).

4.13.1 Resolution by the Telescope

Mark two short, thick black lines about 10 mm apart on a card which is well illuminated and place it 20 m away. The lines make an angle with the eye of

$$\frac{10}{20000} \text{ radians} = \frac{573}{20000} \times 60' = 1.7' \text{ of arc}$$

This is near the limit at which the eye can distinguish one line from the other and is known as the resolving power of the eye.

In a telescope, the diameter **D** of the object lens not only determines the limiting magnitude of the faintest stars we can see, but also determines whether or not we can separate or resolve two stars that are close together, i.e., it determines its resolving power.

The resolving power of a telescope for yellow light can be expressed as an angle α by the simplified expression as

$$\alpha = \frac{138}{D} \text{ in seconds of arc} = 0.92'' ,$$

where α is the smallest angle that can be resolved by the telescope's primary lens.

This is the separation angle for **D** = 150 mm and yellow light of wavelength 550 nanometres (550×10^{-9} m).

This separation, although produced at the primary image, cannot be perceived or appreciated unless magnified by the eyepiece to bring the angle up to 1.5' (minutes of arc which is about the limit of resolution by the normal eye). Magnification at the eyepiece must therefore be at least

$$M = \frac{1.5 \times 60''}{0.92''} = 97.8$$

in order for the separation to be seen by the eye, (see section 4.19).

4.14 THE BARLOW LENS

Many telescope users avoid Barlow lenses as they require some skill in using them, and they are often not well understood. A Barlow lens can, however, be satisfactorily employed in observing, to increase magnification at the eyepiece or to produce an amplified image at Q on a photographic film. A sketch will explain how it works as an accessory of a telescope, Fig. 4.14.1.

B is the Barlow lens which is simply a concave diverging lens of focal length **f**. This is interposed a distance **u** from the primary focus **P** of the object lens. This, in effect, extends the focus from **P** to **Q**. By the lens formula,

$$\frac{1}{u} + \frac{1}{v} = \frac{1}{f}$$

so **v**, the distance **RQ**, equals

$$\frac{uf}{u-f}$$

Note that **u** and **f** are negative when we put numbers in formula.

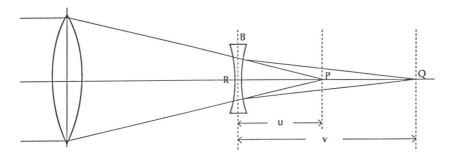

Fig. 4.14.1. How a Barlow lens works. The focal length of the Barlow lens is f the basic relation is

$$\frac{1}{u}+\frac{1}{v}=\frac{1}{f}$$

and the amplification factor equals

$$\frac{v}{u}=\frac{f}{u-f}$$

This is the Barlow lens magnification, or amplification factor. Barlow lenses generally have engraved on them a modest figure of either ×2 or ×3 for their amplification factor. This is because commercially available lenses are fixed within a standard tube to produce one or other of these amplifications. This is, for some, a convenience, but it denies enterprising observers the fun of trying out the lens in all possible positions, because it is fixed in the tube, (see Fig. 4.14.2).

The amplification factor is

$$A = \frac{f}{u-f}$$

and **A** becomes infinite when **u = f**.

For normal use, where say for example, **f** = 75 mm and **u** = 45 mm, so

$$A = \frac{75}{30} = 2.5$$

But if $\mathbf{u} = 60$, then

$$A = \frac{75}{75 - 60} = \frac{75}{15} = 5$$

A little enterprise will tempt you to make **u** greater than **f**. This completely changes the situation, and the sign, and turns the arrangement, in principle, to a Galilean telescope with an upright virtual image formed between the object lens and the Barlow lens eyepiece. This principle is used in opera glasses. This type of Galilean telescope is of little use in astronomy as it produces a magnification of only about 3 or 4. It is however used in terrestrial observing (for instance, bird watching) as then the image is upright.

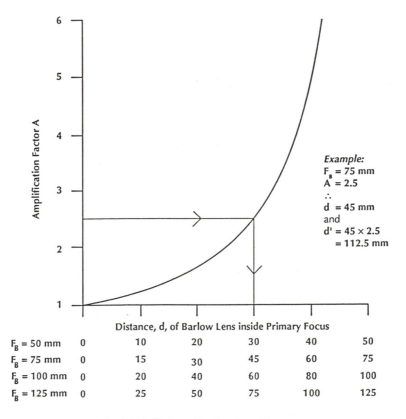

Fig. 4.14.2. Universal Barlow Lens Diagram.
The diagram shows how A, the amplification factor, varies with d, the distance of the Barlow Lens inside the Primary Focus, for different values of F_B the Barlow lens focal length. Note that d, the Primary Focus and the New Barlow Focus, is given by the formula, $d' = Ad$. Designed for the Journal of the British Astronomical Association by Commander H. R. Harfield, R.N. (Rtd).

4.15   OBSERVING WITH EASE AND COMFORT USING A PAIR OF BINOCULARS OR A SMALL TELESCOPE

Binoculars are widely acclaimed as an ideal instrument for sky watchers. They are convenient for looking at small star groups, comets or galaxies as they have a wide field of view, of about 5°, that enables observers to recognise star configurations without getting lost among the stars. They have, however, three serious draw-backs.

(1)   They are impossible to hold still.
(2)   They are uncomfortable to use on objects that are more than about 20° above the horizon and tend to give the ardent observer a pain in the back or neck.
(3)   They normally have no means of indicating or measuring altitudes or azimuths of celestial bodies.

Fig. 4.15.1. The D.I.Y. Binocular Stand in use

These three disadvantages can easily and satisfactorily be overcome by the use of the D.I.Y. Binocular Stand shown in the photograph Fig. 4.15.1 and described by the Author in *The Journal of the British Astronomical Association* (Vol 92, No 3, April 1982). The principles involved are as given earlier in this book in Fig. 1.10.5.

Fig. 4.15.2, on the next page, illustrates the way in which binoculars are supported on Stand **A** holding them at an angle of 50° to the horizontal platform **B** which is free to turn about a central bolt through the centre of an azimuth table **C** marked in degrees of azimuth. In use the azimuth circle is oriented and fixed so that its North-South line is in the meridian. The pointer **D** is attached to the free-to-turn base **B** and is in line with the optical axis of the binoculars. So **D** shows the *azimuth* of the line of sight of the binoculars.

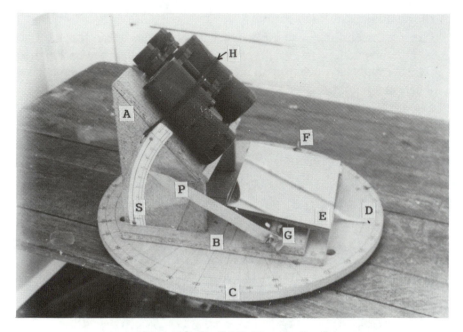

Fig. 4.15.2. The D.I.Y. Binocular Stand.

The small platform shown supporting a telescope at an open window which appears in Fig. 4.15.3 can be a great convenience as it transforms a living room or bedroom into an improvised miniature observatory. It thus relieves the sky watcher of the inconvenience of venturing outdoors at night. The platform can be used for steadily supporting binoculars, small telescopesas in Fig. 4.15.3, or a camera, either for short exposures as in Fig. 4.15.4 or for photographing star trails as shown in Fig. 1.12.4.

The swivel mirror device used with binoculars can very successfully be applied to a small 60 mm telescope mounted as shown in the photogragh (Fig. 4.15.3). It enables the observer to place the telescope on a small table by an open window and to observe stars from a few degrees above the horizon, to altitudes of 80° without moving the actual telescope.

In Fig. 4.15.2, firstly the base **B** is turned to bring **D** to be in the vertical plane of the object to be observed.

Secondly, the binoculars having been set for the object's *azimuth*, the mirror **E** now deals with the object's *altitude*. **E** is a plane mirror, about 200 mm × 150 mm, mounted with a good adhesive on this axis held in bearings at **F** and **G** which are two tool clips.

Fig. 4.15.3. The reflecting swivel mirror device in use with a small 60 mm telescope.

The mirror is then turned about its axis **FG** until the object selected for observation appears in the eyepiece of the binoculars. The optics are simple and instructive. For example, a star at an altitude of 50° will be in view when the mirror surface is horizontal (check this with a spirit level).

The angle of incidence = the angle of reflection (θ).

So by simply turning the mirror, the observer can scan objects from the horizon to the zenith comfortably seated and without touching the telescope which remains firm and fixed. An arm **P** (pointer) fixed by nut and bolt to the axis at **G** passes over a scale calibrated so as to give the altitude in degrees. The device thus turns a pair of binoculars with their disadvantages (which we listed above) into a steady and useful instrument that can measure the altitudes and azimuths of stars to within a degree. This valuable facility can then be used in conjunction with the nomograms or a planisphere with an alt–Az graticule, to identify celestial objects by converting their alt–Az into their corresponding Right Ascension and Declination, given, for example, in the *Handbook of the British Astronomical Association.*

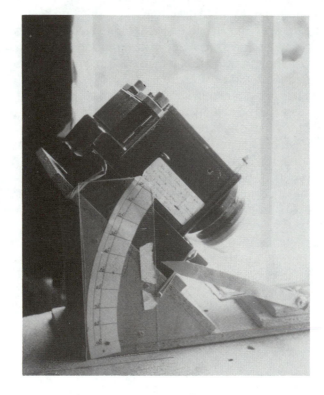

Fig. 4.15.4. The reflecting swivel mirror device in use with a camera that can conveniently be brought to bear on any part of the sky, simply by moving the base in Azimuth, and by moving the mirror arm for the altitutde as registered by the altitude pointer.

The platform **A** in the figure can satisfactorily be used to hold a small refractor telescope at 50° to the horizontal using two cup hooks and a couple of strong elastic bands (see Fig. 4.15.3).

It should be noted that a cheap plane mirror can be used for merely finding the positions of stars, or for identifying them. For satisfactory observing, the mirror used in this device should be optically plane "top silvered" to avoid the problem of refraction mentioned earlier in Section 4.1.1.

4.16   RAINBOWS

One of the many exciting things that the Sun does to brighten our lives is to produce rainbows. As the poet, Wordsworth, wrote:

   "My heart leaps up when I behold a rainbow in the sky".

Rainbows are formed when the Sun shines behind us and we are lucky enough to be facing a rain cloud that is raining down raindrops profusely. This produces a lovely arc of colour in the sky which encircles the rain cloud.

If you measure the arc, with an angle measuring device, the arc will be found to be 82° across its diameter. It will be about 4° in width and will literally contain "all the colours of the rainbow", with the outer edge red and the inner rim violet, which are caused by refractions by the water drops.

Here is something to do to simulate a rainbow, and its colours. Fill a garden spray-gun with water, and on a sunny day stand with your back to the Sun and send a fine spray of water into the air where the shadow of your head is situated. Round about you is a miniature rainbow 82° across. To understand how this comes about, consider a single raindrop in the Sunlight. This can be simulated by filling a spherical flask (borrowed, perhaps, from the Chemistry Lab) and then hold it up in the Sun and place a white cardboard screen, which has a hole cut in it of the same diameter as the flask, a few centimetres in front of the flask, between it and the Sun.

A beam of Sunlight passes through the hole in the cardboard and the top edge of the beam enters the spherical flask at **B** and emerges after refraction at **B**, reflection at **C** and another refraction to at **D** and appears on the screen at **E** as a small arc, red outside and violet inside (as shown in Fig. 4.16.1 and Fig. 4.16.2). **DE** is about 40° to the Sun line **AB**. The same happens to all edges of the incident beam with the result that on the screen is a complete circular spectrum encircling the hole. For the mathematician and physicist, the angles involved can be calculated without great difficulty, as is shown in the next Section.

4.16.1   Diagram to explain the formation of a rainbow
Light from the Sun is incident on the spherical flask in the direction **AB**. It is refracted at **B**, reflected at **C**, and again refracted at **D** to emerge in the direction **DE** (see Fig. 4.16.2). The angle between **AB** and **DE** for red light is 42.8°, and is 40.8° for violet light giving a rainbow with red at the outside edge about 2° in width. As there is a large volume of raindrops, this deviation angle of 41° takes place over a

large area and so we get the approximate mean angle **AFE** of 41° for a circle round the original direction of the Sun, having a diameter of $2 \times 41° = 82°$.

Fig. 4.16.1. Creating a Rainbow using a spherical flask of water to simulate a rain-drop.

In physical optics, we learn by geometry and by Snell's Law which states the refractive index of water,

$$u = \frac{\text{Sin } i}{\text{Sin } r} = 1.329 \text{ for red light.}$$

It can be shown that

$$\text{Cos } i = \sqrt{\frac{u^2 - 1}{3}}$$

and if we use μ for red light, 1.329 then **i** = 59.6° and **r** = 40.5°, so that the deviation $4r - 2i = 42.8°$.

If we use μ for violet light, 1.343, then $\mathbf{i} = 58.8°$ and $\mathbf{r} = 39.6°$, giving the deviation as 40.8°. Consequently the width of the rainbow is $42.8° - 40.8° = 2°$ between red and violet.

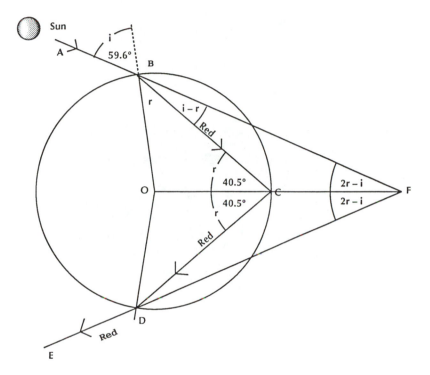

Fig. 4.16.2. Diagram to explain the formation of a Rainbow. Showing only the Red rays.

4.17 LIGHT INTRUSION THE BANE OF SKYWATCHERS

Sky watchers are often annoyed or frustrated because, in towns or near motorways, their lights interfere with observations that require a dark background, particularly if astrophotography is to be attempted. There is only one thing to be done under present conditions.

Take binoculars, or a small telescope, mounted for mirror reflection as shown in the Figs in Section 4.15, to a spot away from city lights in a car. Place it firmly on a support across the passenger seat and use it sitting in the driver's seat in comfort. The car is parked suitably for the part of the sky required and the mirror provides the variation in altitude from $0°$ to almost overhead, in the same way as the device can be used at an open window of a room. To keep the car steady a couple of car jacks are very useful.

Fig. 4.17.1. Observing away from city lights.

4.18   SPECTRA OF STARS

The other possible activity in an atmosphere pervaded with neon lights, many of which are mercury vapour or sodium lights, is to use them for practice to represent starlight from stars, and photograph or study visually the spectra of the various lighting systems using a good prism. Place the prism just in front of the object lens, or use a special grating or replica which will produce spectra and demonstrate spectral lines, which play such an important part in astronomy.

Fig. 4.18.1 shows a camera mounted to take photographs of the spectra of a bright star in the line of sight of the tube **DC**. The camera is held at **B** and the dispersed spectra from the grating at **A** is received in the camera lens **E**.

The mounting is of interest as it shows the azimuth table which holds the camera, and the altitude of the object is measured by the protractor and plumb-line.

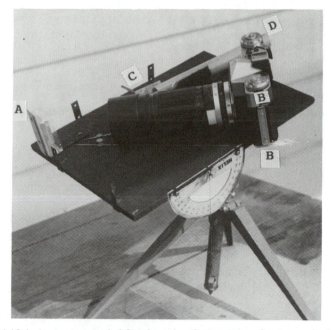

Fig. 4.18.1. Apparatus needed for observing the spectrum of a star or city lights.

With care the spectrum of a star of the first magnitude can be photographed using a grating as shown at **A** and a camera at **B**. **CD** is a sighting tube to help in aligning the device on the star. The platform supporting the camera and grating is mounted on a tripod stand fitted with a protractor and plumb-line to give an approximate measurement of the star's altitude. Ideally the platform should be mounted on an equatorial telescope driven by sidereal clockwork, for exposures of more than a few seconds.

Particulars of the grating used may be of interest to physics students. The grating has 1200 grooves per mm, spacing distance 8.333×10^{-4} mm. θ the angle made by a wavelength is given by $\lambda / d = \sin \theta$

For red light
$\lambda = 655$ nanometres, $\theta = 52°$ $\sin \theta = \dfrac{655 \times 10^{-9}}{8.33 \times 10^{-7}} = 0.786$

For yellow light
$\lambda = 536$ nanometres, $\theta = 40°$
For violet light
$\lambda = 415$ nanometres, $\theta = 30°$

These figures indicate the dispersion made by the grating.

4.19   WHAT A TELESCOPE CAN DO UNDER IDEAL CONDITIONS AND RELATED FORMULAE

Function	Description	Depends on
Light Gathering Power	L.G.P. Section 4.3.1	The diameter of Objective D only (in mm)
Magnification	for complete telescope, M Object part = M_o Eyepiece part = M_e $M = M_o \times M_e$ = Magnification at primary focus as seen by the unaided eye Section 4.3.3	F = focal length f of object f = focal length of eyepiece S = distance of distinct vision (for normal eye, 250 mm)
Magnification in Photography	By eyepiece projection Section 4.11	F, f, and v v = distance of film or plate from eyepiece
Field of View, d	If t = time in secs for star declination δ to cross a field of diameter d. Then d in arc mins $= \dfrac{t \cos \delta}{4}$ Section 4.12	Magnification of telescope M and field of view of eyepiece
Resolution, Resolving Power	Smallest angle in arc secs that can be resolved at the primary focus (α) Section 4.13.1	α depends only on D but for visual resolution α must be magnified by the eyepiece to at least 1.5′ and so depends on f as well.
Barlow Lens	Amplifying factor = M_B Section 4.14	f_B is the focal length of a Barlow Lens., u is the distance between the Barlow lens and the primary image

Theoretical Relations

Night adapted eye pupil $\approx$ 8 mm diameter LGP over unaided eye $= \dfrac{D^2}{S^2}$

Limiting Magnitude M of a telescope is M = 2 + 5 $\log_{10}$ D (for D = 100 M = 12)

Approximately $M = \dfrac{F}{f} = \dfrac{D}{d}$ where d = diameter of exit pupil of eye, and is true only if the eye is focussed on image at ∞

M_o = magnification of distant object with image at primary focus as seen by normal eye at 250 mm $= \dfrac{F}{250}$

M_e = magnification by the eyepiece $= \left(\dfrac{250}{f} + 1\right)$

Total Magnification $M = M_o \times M_e = \dfrac{F}{250}\left(\dfrac{250}{f} + 1\right) = \dfrac{F}{f} + \dfrac{F}{250}$

Magnification of primary image by eyepiece $M_P = \left(\dfrac{v}{f} - 1\right)$ the Effective Focal Length

$= F\left(\dfrac{v}{f} - 1\right)$ and the Effective Focal Ratio $= \dfrac{F}{D}\left(\dfrac{v}{f} - 1\right)$

Field of view of sky by an observer at the telescope =
$\dfrac{\text{Field of view presented by eye piece}}{M} \approx \dfrac{40}{M}$

(Field of view of eyepieces may range from 35É to 45É)

Resolving Power in theory $= \dfrac{\lambda}{D}$ but in practice is more accurately expressed as

$\dfrac{1.22\,\lambda}{D}$ Radians (λ for yellow light = 550 nm). Resolving Power $= \dfrac{138}{D}$ in arc secs

(For D = 150 mm, α = 0.92"For eye resolution use eyepiece magnification

$\dfrac{1.5 \times 60}{0.92} = 97.8$ and f must not be more than 2.5 mm

$M_B = \dfrac{f_B}{f_B - \mu}$ Effective focal length of a telescope with a Barlow lens $= \dfrac{F\,f_B}{f_B - \mu}$

F is the focal length of telescope objective.

5

Miscellaneous calculations

In the foregoing chapters we have seen how astronomy can provide numerous activities for skywatchers and students, particularly those who are enterprising enough to use a calculator or simple workshop tools to make D.I.Y. devices for observing or measuring. This chapter is intended to supplement some of the earlier sections in this book with examples from astronomy, space travel and mathematics, with the object of integrating these subjects when appropriate into the science curriculum. Beginners in astronomy are often convinced that mathematics is a kind of ploy devised to undermine their self-confidence instead of being a means that can lead them to a proper understanding and enjoyment of their subject. This is particularly true now that scientific calculators and personal computers have eliminated so very much of the many tedious calculations that were required in the days of logarithms, slide-rules and trigonometrical tables.

5.1 THE UNIVERSAL LAW OF GRAVITY

In Chapter 2 (Section 2.1.1), Newton's Law of Gravity was considered as the universal law controlling the motions of the entire solar system, but it is of interest to apply the law to attractions between familiar objects on Earth.

Two cricket balls just touching experience a force of attraction of 3.27×10^{-10} Newtons. The mass of a cricket ball is 16.16 grams and its radius is 3.65 cm. This information although of some interest will not be of any use in the game of cricket, but if we consider the attraction between two large tankers lying side by side the attractive force is well within our daily experience as the following calculations using Newton's Law of gravitation will show, assuming that the masses of the tankers behave gravitationally as if their masses can be considered as acting at their centres of gravity. Suppose the loaded mass of each vessel to be 330 000 tonnes, and when lying along-side each other centres of mass are 33 m apart; then using these values the force of attraction between the two ships will be

$$\frac{G\,M_1 \times M_2}{33^2} \text{ where } M_1 = M_2 = 3.3 \times 10^5 \text{ tonnes}$$

$$= \frac{(3.3 \times 10^5)^2 \times 10^6 \times 6.67 \times 10^{-11}}{33^2}$$

$$= 6.67 \times 10^3 \text{ newtons (approx.)}$$

$$= 667 \text{ kg Wt}$$

$$= 0.667 \text{ tonnes Wt.}$$

The force of attraction is approximately 0.667 tonnes weight, and dockworkers handling large ships are aware of the possible danger of being trapped between two ships lying side by side (see Fig. 5.1).

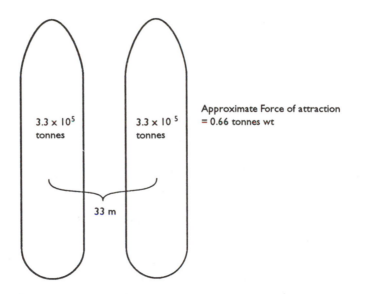

Fig. 5.1. The approximate force of attaction between two large oil tankers.

5.2   GRAVITY AND SHIPS SAILING AT SEA

A ship of 20 000 tonnes sailing due East along the equator at a speed of 20 knots then turns about and sails due West at the same speed.

   This change in direction results in a small change in the ship's weight, because of the small change in the centrapetal force acting on the ship.

A speed of 20 knots is approximately 10 metres per second. The ship when apparently at rest on the sea at the equator is actually moving round the Earth's axis at a speed of 2π r in 24 hours, where r is the radius of the Earth, 6.378×10^6 m. This speed is thus

$$\frac{2\pi \times 6.378 \times 10^6}{24 \times 60 \times 60} = 464 \text{ m/sec.}$$

and the centrapetal acceleration is given by

$$\frac{v^2}{r}.$$

When sailing due East V is incresed to V + 10, but when sailing due West, V is decreased to V − 10, and the centrapetal accelerations become

$$\frac{(V + 10)^2}{r}$$

and

$$\frac{(V - 10)^2}{r}$$

respectively.

The difference between these two values is

$$\frac{4 \times 10 \times V}{r} = \frac{40 \times 464}{6.378 \times 10^6} \text{ ms}^{-2} = 2.91 \times 10^{-3} \text{ ms}^{-2}$$

This change in the centralpetal acceleration on the ships mass of 20 000 tonnes, produces a change in the ship's weight (20 000 000 Kilograms), given by

$$2 \times 10^7 \times 2.91 \times 10^{-3} = 5.82 \text{ Tonnes weight.}$$

This represents a change of 0.0291%.

It is of interest to apply the principle above to an athlete of say mass 70 kg. This will show that the althlete has a slightly better chance of breaking his personal record by runing in an Easterly direction instead of running in a Westerly direction. By doing this his weight, (but not his mass) will be made less by about 20 grams. The reduction is in effect equivalent to a reduction of the force of gravity by about 0.0291%. It will be found that the effects discussed above, must be modified for latitudes that are not on the equator, and for latitude ϕ the athlete experiences a loss of weight of 20 cos ϕ grams wt.

5.3   THE PENDULUM

Pendulum studies form a part of a physics course and were used by Galileo, the physicist and astronomer, during his enquiries into force, mass and motion and led to the relation for small amplitudes

$$T = 2\pi\sqrt{\frac{l}{g}}$$

for the time of oscillation of a pendulum of length l and acceleration due to gravity g. It was soon discovered that g varied slightly with the position of the pendulum on the Earth's surface. The Earth is flattened at the poles and so the polar pendulum is a little nearer the Earth's centre, with gravity a little stronger, and oscillates a little more quickly than it would do at the equator. The polar radius

$$= 6.356775 \times 10^6 \text{ m}$$

and the equatorial radius

$$= 6.378160 \times 10^6 \text{ m.}$$

Another influence on gravity (see Section 5.2) is the fact that the Earth is spinning and so a person of mass m at the equator moving with velocity v weighs less by

$$\frac{mv^2}{r}$$

The two effects together will cause the pendulum that oscillates, say, with a period of 1 second at the equator, to oscillate a little quicker at the North Pole, and so gain about 4 minutes each day. By sheer coincidence a normal clock pendulum keeping G.M.T. at the equator will keep Sidereal Time at the pole. This is an interesting calculation, using

$$T = 2\pi\sqrt{\frac{l}{g}}$$

and taking into account the two effects described above.

5.4 THE FOUCAULT PENDULUM

In 1851, Foucault carried out an experiment which can be repeated in any room with a high ceiling, to demonstrate that the Earth is rotating about a North–South axis. A pendulum consisting of a heavy steel ball is suspended by a fine wire from the roof of a high building, and set in motion by being drawn to one side by a thread, which, when all is steady, is burnt to set the pendulum in an oscillatory motion in a vertical plane. As the pendulum is long and heavy it will continue to oscillate for several hours. It will be observed that the plane of the oscillation appears to change its direction relative to the objects in the room but its plane of oscillation remains fixed in space relative to the stars. The change in direction of swing relative to the room is due to turning of the Earth, which is not fixed in space. It will be seen that at the equator with the pendulum ball starting its swinging in the plane of the equator, it will continue to swing in that plane (see Fig.

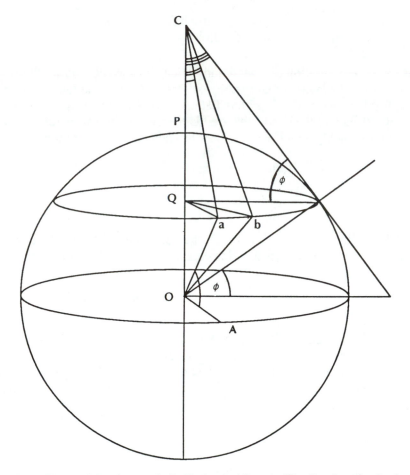

Fig. 5.4.1. The pendulum is set swinging in the meridian at a. The direction of swing is along the tangent aC to the Earth at the point a, which meets the axis OP produced at C. After a time the Earth will move through an angle from West to East and take the pendulum to b but keeping it on the same latitude. The pendulum still swings in space in the dirction ac. So relative to the ground it has changed its direction through an angle from bCa to bQa.

$$\frac{\text{Rate of rotation of pendulum}}{\text{Rate of rotation of the Earth}} = \frac{ab}{aC} \div \frac{ab}{aQ} = \frac{Qa}{aC}$$

but

$$\frac{Qa}{aC} = \sin aCQ = \sin aOA = \sin \phi$$

where ϕ = Latitude of a

So the plane of oscillation of a Foucault pendulum in Latitude ϕ turns through $15°\sin \phi$ each Sidereal hour or during a Sidereal day it turns through $360°\sin \phi$.

5.4.1). There can be no influence to cause it to change this plane of swinging so it continues to swing in an East–West direction and there is no apparent rotation of this direction of swing.

Suppose you could set up your pendulum at the North Pole. It will continue to swing in a constant plane with respect to the stars, but the Earth beneath the pendulum would be turning anti-clockwise with respect to the stars at the rate of $15°$ per hour. So for an observer at the North Pole the pendulum swing will appear to follow the stars and change its Earth direction of swing steadily clockwise at $15°$ per hour. At a latitude ϕ between the equator and the North Pole, it can be shown (see Fig. 5.4.1) that the direction of the change in the direction of swing is clockwise, but its progress will be $15° \sin \phi$ per hour, i.e., when $\phi = 0$ there is no change and when $\phi = 90$ at the Pole the change in direction is at the rate of $15°$ per sidereal hour.

This experiment carefully conducted could be a means of finding your latitude if you landed on a desert island with an overcast sky, with only a watch or radio set to mark the hours!

5.5   METEORS AND SHOOTING STARS

Sky watchers and indeed anyone in open country on a dark clear night are intrigued by the streaks of light of meteors or shooting stars that occasionally appear. These are caused by small particles of interplanetary dust entering the Earth's upper atmosphere at such a high speed that they become white hot through air friction and leave a trail of incandescence.

Meteor particles often occur in swarms which travel around the Sun in elliptical orbits and some such swarms have been identified as the remnants of comets. Consequently, whenever the Earth's orbit passes through such a swarm of particles 'Shooting stars' appear. They seem to radiate from a particular part of the sky on particular dates. For example, in mid-August the 'radiant' is in Perseus and the shower is listed in astronomical handbooks and year books as the Perseus Shower or Perseids.

It happens from time to time that particles much larger than particles of dust or grains of sand enter the Earth's atmosphere and are sufficiently large to withstand the 'burn up' that small particles suffer, and hit the Earth's surface with violence and danger. Bodies that reach the Earth are known as meteorites. These can vary in size and mass from a few milligrams to thousands of kilograms. The former may fall on the Earth or into the sea, and are scarcely noticed, but in the past meteorites weighing millions of tonnes have fallen on to the Earth causing widespread damage, deep craters, and sending thick columns of dust and steam into the atmosphere, resulting in abrupt changes in our climate and the destruction of established species of animals. Such calamities are recorded in the fossil remains of thousands of millions of years ago now found in rocks and river beds. The Arizona Meteorite Crater is perhaps the most outstanding example of such a calamity.

5.5.1 Simple circular orbits

Consider a satellite of mass, m, in circular orbit round the Earth with a velocity, v, at a distance from the centre of the Earth of R. The mass of the Earth is M.

The acceleration of the satellite towards the centre of the Earth is v^2/R, and this is caused by the gravitational attraction of the Earth which is given by the universal Law of Gravity

$$F = \frac{GmM}{R^2}$$

and this = m x acceleration towards the centre. We have therefore,

$$\frac{mv^2}{R} = \frac{GmM}{R^2} \tag{1}$$

where G is the gravitational constant.
If the periodic time of the orbit is T, then

$$vT = 2\pi R$$

or

$$T = \frac{2\pi R}{v}$$

or

$$v = \frac{2\pi R}{T}$$

Substituting the value of v in equation [1] we have,

$$\frac{m^4\pi^2 R^2}{T^2 R}$$

or

$$T^2 = \frac{(2\pi)^2 R^3}{GM}$$

It is convenient to consider R as being equal to r + h where r is the radius of the Earth and h the height of the satellite above the Earth's surface. Therefore,

$$T = 2\pi\sqrt{\frac{R^3}{GM}} \tag{2}$$

and T^2 is thus proportional to the 3rd power of distance (r + h).

$T^2 \propto (R+h)^3$ which is Kepler's 3rd Law which states that the ratio of the square of the period to the cube of the semi major axis is constant. The semi major axis of an ellipse is the radius of the orbit for a circular orbit.

5.5.2   Determining the geosynchronous orbit

A practical problem of current interest suggests itself, namely, at what height above the Earth should a satellite be orbiting in order that the time of its orbit would be 24 hours? What is the value of h for a geosynchronous orbit, which is now used in television systems, when an apparently permanent satellite over a particular region is required?

We have from equation [2] above

$$R^3 = \frac{T^2 MG}{4\pi^2}$$

where

$$T = 24 \times 60 \times 60 \text{ sec,}$$

$$G = 6.670 \times 10^{-11} \text{ kg}^{-1} \text{ m}^3 \text{ s}^{-2}$$

$$M = 5.98 \times 10^{24} \text{ kg}$$

$$R^3 = \frac{(24 \times 3600)^2 \times 6.670 \times 10^{-1} \times 5.98 \times 10^{24}}{4\pi^2}$$

$$= \frac{(2.4 \times 3.6 \times 10^4)^2 \times 6.670 \times 10^{-11} \times 5.98 \times 10^{24}}{4\pi^2}$$

$$R^3 = \frac{2977.52}{4\pi^2} \times 10^{21}$$

$$= 75.42 \times 10^{21} \text{ m}$$

$$R = 4.225 \times 10^7 \text{ m}$$

The equatorial radius pf the Earth, r, is

$$6.378 \times 10^6 \text{ m}$$

But

$$R = r + h$$

$$= 4.225 \times 10^7$$

$$= 42.25 \times 10^6 \text{ m}$$

but subtracting the radius of the Earth

$$r = 6.38 \times 10^6 \, m$$

$$h = (R - r)$$

$$= 35.87 \times 106 \, m.$$

Therefore h (height above the surface of the Earth) is 35.87×10^6 m or, h = 35,870 kilometres which is the recognised approximate position for a geosynchronous satellite, now extensively used for radio and television transmission.

We can calculate the velocity of escape of a body from the Earth, from energy considerations.

Suppose a 1 kg mass to be at a distance r from the centre of the Earth. The force of attraction on the mass is

$$\frac{G \times \text{Mass of the Earth} \times 1}{r^2}$$

where G is the Gravitational Constant.

If the 1 kg mass falls a small distance dr towards the Earth, it loses a little potential energy, dw, which increases its kinetic energy by

$$dw = \frac{GM \times 1}{r^2} \, dr.$$

The kinetic energy acquired in falling from infinity to Earth is

$$W = \int_0^R \frac{GM}{r^2} \, dr = \frac{Gm}{R} \, .$$

$$= \frac{6.670 \times 10^{-11} \times 5.98 \times 10^{24}}{6.37 \times 10^7} \, \text{Joules}$$

$$= 6.261 \times 10^7 \, J \, .$$

This energy is released at the Earth's surface as kinetic energy of the 1 kg mass having velocity v i.e. $\frac{1}{2} \times 1 \times v^2$ so $\frac{1}{2} \times v^2 = 6.261 \times 10^7$ and $v = 11.2 \times 10^3 ms^{-1}$.

This velocity of a body on reaching the Earth after falling from a great distance, can be used to find the effect that a massive body such as an asteroid would have on striking the Earth. During the past few million years there is evidence that the Earth has suffered major calamities due to impacts from asteroids of various sizes. These impacts may have caused changes in our atmosphere and climate that resulted in the extinction of biological species including the dinosaurs.

The damage done by the impact of a small asteroid can be roughly estimated by considering the kinetic energy of the body being transformed into heat. First, for simplification consider the effect of one kilogram of ice at 0°C hitting the Earth with a velocity equal to its escape velocity 11.2 kg per second. The kinetic energy of this kilogram is $\frac{1}{2}mv^2$ which is

$$\frac{1}{2} \times (11.2 \times 10^3)^2 \text{ Joules.}$$

This energy is transformed into heat and in effect would raise the temperature of the kilogram mass by T degrees, in accordance with the equation

$$\frac{1}{2}(11.2 \times 10^3)^2 = 4.2 \times 10^3 T$$

giving

$$T = 14900°C.$$

The amount of heat developed could raise the temperature of 149 kilograms of water from 0° to 100°C i.e. to boiling point. Now instead of 1 kg of water in the form of a solid striking the Earth at a speed of 11.2 km.s^{-1} consider an actual typical asteroid sized mass of ice and suppose it were to fall in the Pacific Ocean. The energy of this impact would raise the temperature of the surrounding waters of the ocean, as the following calculation will show. Suppose the average depth of the world's oceans to be 1.5 km and that this covers 2/3 of the Earth's surface. This is enough water to cover the Earth's surface to a depth of about 1 km, and the volume of this water mantle is

$$4\pi r^2 \times 10^3 \text{ m}^3$$

where r is the radius of the Earth

$$6.378 \times 10^6 \text{ m.}$$

The amount of water is thus very roughly

$$5.11186 \times 10^{17} \text{ m}^3.$$

This is a small fraction of the total volume of the Earth.

To raise the temperature of this quantity of water by 100° we require an ice asteroid of volume 1/149th of this which is

$$\frac{5.11186 \times 10^{17}}{149} = 3.43 \times 10^{15} \text{ m}^3$$

This radius of such a spherical body would be r where

$$\frac{4}{3}\pi r^3 = 3.43 \times 10^{15}$$

giving the radius of the asteroid r = 93.5 km.

If an asteroid of ice of this size were to strike the Earth it would have sufficient energy to bring our oceans to the boil.

5.5.3   A putative tunnel

As the Channel Tunnel is now complete, I am reminded of a typical problem that was published in the Newsletter of the Association for Astronomy Education (January 1987 and April 1987). This provided a good mathematical exercise as it involves our planet, the force of gravity, simple harmonic motion, and the velocity of an artificial satellite in orbit round the Earth.

Suppose a tunnel could be drilled through the centre of the Earth connecting Britain to New Zealand. Now consider the motion of a small mass, m, dropped down the tunnel, (ignoring the hazards of high temperature and friction and assuming the Earth to be a uniform sphere) how long will it take for the mass to reach New Zealand?

In Fig. 5.5.3 ADB represents the Earth of mass M. Its centre is at O and AOB is the tunnel. When the body is at A the force on it is

$$\frac{GMm}{R^2}$$

and when at C the force is

$$\frac{GM_1 m}{r^2}$$

where M_1 is the mass of the shaded sphere CFE. The unshaded hollow sphere has no gravitational field or effect on M_0 and

$$\frac{M_1}{M} = \frac{r^3}{R^3} \; .$$

The force on M at C is therefore

$$\frac{GMm \, r^3}{r^2 R^3} = \frac{GMmr}{R^3}$$

using the relationship that

$$\text{Force} = \text{Mass} \times \text{Acceleration}$$

$$\frac{M d^2 r}{dt^2} = - \frac{Gm}{R^3} \cdot r$$

This represents the Simple Harmonic Motion of period T

$$T = 2\pi \sqrt{\frac{R^3}{GM}}$$

So the mass m accelerates to the centre where it would reach its maximum velocity and come to rest at B and then return to A.

Using approximate values, we have the time for a *complete* oscillation

$$T = 2\pi \sqrt{\frac{(6.36 \times 10^6)^3}{6.67 \times 10^{-11} \times 5.976 \times 10^{24}}} = 84.13 \text{ minutes.}$$

So, Time to go through the tunnel from A to B = $42^m\ 8^s$.

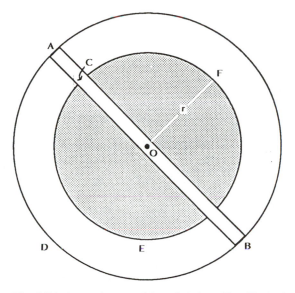

Fig. 5.5.3. A putative tunnel from Britain to New Zealand.

In Section 5.5.1 equation [2] it was found that a satellite encircling the Earth just above the surface of the Earth did so in Time T given by

$$T = 2\pi \sqrt{\frac{R^3}{GM}}$$

which is the same as the period of the mass in the tunnel for its SHM. The oscillatory motion of the mass in the tunnel is the projection of the satellite motion on to the straight line of the tunnel. The maximum velocity reached at the centre of the Earth, O, is the same as the constant velocity of the satellite.

If this velocity is v then acceleration towards the centre is given by

$$\frac{mv^2}{R} = \frac{GmM}{R^2}$$

The centripetal force = gravitational attraction.

$$v^2 = \frac{G \cdot M}{R}$$

$$= \frac{6.67 \times 10^{-11} \times 6 \times 10^{24}}{6.36 \times 10^6}$$

$$= 6.292 \times 10^7$$

and

$$v = 7.932 \times 10^3 \text{ ms}^{-1}.$$

It orbits the Earth in 84.13 minutes. It travels

$$7.932 \times 10^3 \times 84.13 \times 60 \text{ m}$$

$$= 4.0039 \times 10^7 \text{ m}$$

which corresponds to the circumference of the Earth

$$= 2\pi \times 6.36 \times 10^6 \text{ m}$$

$$= 4 \times 10^7 \text{ m}$$

5.5.4   A tunnel between any two points

The problem 5.5.3 invites attention to dropping a smooth object of mass m down a smooth tunnel connecting any two points on the Earth's surface as illustrated in Fig. 5.5.4 by AB. Here the force on mass, m, towards the centre of the Earth

$$= \frac{GmM_1}{r} \qquad\qquad [1]$$

where, M = mass of the Earth and M_1 = Mass of the sphere of that part of the Earth with radius r

Thus as

$$\frac{M}{M_1} = \frac{R^3}{r^3}$$

then

$$M_1 = \frac{M}{R^3} r^3$$

If one substitutes this in Equation [1] above, then the force on mass, m, toward C at E

$$= \frac{Gm}{r^2} \frac{M}{R^3} r^3 \sin \theta$$

Let $r \sin \theta = x$, the distance from the centre point D.

Then the

$$\text{Force at E to C} = Gm\,\frac{M}{R^3}\,r\,\sin\theta$$

$$= Gm\,\frac{M}{R^3}\,x$$

$$m\,\frac{d^2x}{dt^2} = \frac{GmM}{R^3}\,x \quad \text{This represent S.H.M.}$$

and

$$T = 2\pi\sqrt{\frac{R^3}{GM}}$$

which is the same time of oscillation as for m falling along a tunnel along a diameter 84.13 minutes, i.e., 84.13 minutes for a complete oscillation or 42.07 minutes for motion A to B.

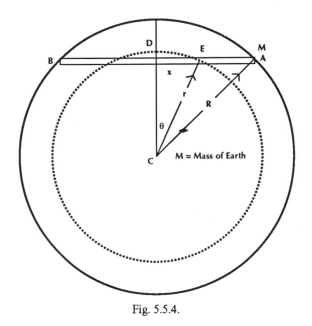

Fig. 5.5.4.

5.5.5 A falling meteorite
Here is a little thought provoking problem that involves physics, gravity, kinetic energy, relative velocities, astronomy and the hour angle of the Sun.

'A meteorite from far beyond the solar system falls vertically down upon the Earth's surface straight down a well. What was the time at which this happened?'

At first glance it may appear to be a trick question, but it requires a little astronomy and dynamics to obtain a plausible answer.

Hints for the working of this problem are as follows, leaving reader to work it out as an exercise in dynamics and astronomy.

The velocity V achieved by a meteorite of mass, m, falling from ∞ towards the Sun, which is of mass M is given by,

$$\frac{1}{2}m\,V^2 = \int^R G\,\frac{mM}{R^2}\,dR$$

where R is the radius of the Earth's orbit round the Sun.

$$V^2 = \frac{2GM}{R} \qquad\qquad [1]$$

The velocity v of the Earth in its orbit round the Sun is determined by the relation

$$\frac{Mv^2}{R} = \frac{GMm}{R^2}$$

giving

$$v^2 = \frac{GM}{R} \qquad\qquad [2]$$

from [1] and [2]

$$V^2 = 2v^2$$

and

$$\frac{V}{v} = \sqrt{2}$$

This makes the *Relative* velocity of the asteroid, *observed* to go straight down the well to be moving in fact at an angle to the vertical,

$$\arctan \frac{1}{\sqrt{2}} = 35°.26$$

This angle is thus the Hour Angle of the Sun, so the time of the occurrence is 2^h 21^m past midnight.

Comets by virtue of their composition and gradual disintegration leave a trail of dust and debris behind them which is largely responsible for meteor showers. Halley's Comet, for example, is associated with the October Orionids shower.

5.5.6   The Earth–Moon barycentre

Two masses M and m that under gravitational attraction such as the Earth and its Moon, revolve round their common centre of gravity v (the barycentre) which is situated at a distance r from the centre of the Earth, and R from the centre of the moon. Fig. 5.5.6.

Then

$$Mr = mR$$

and

$$\frac{R}{r} = \frac{M}{m}$$

R + r = distance between the centres of Earth and Moon

$$= 3.844 \times 10^8 \text{ m} \qquad\qquad [1]$$

and as the ratio

$$\frac{M}{m}$$

is known to be

$$\frac{81}{1} = \frac{R}{r} \qquad\qquad [2]$$

from [1] and [2]

$$r = 4.7 \times 10^6 \text{ m}.$$

So r is less than the Earth's radius and the centre of gravity of the Earth–Moon system is $(6.378 - 4.7)\ 10^6$ m inside the surface of the Earth, or 1.678×10^3 kilometres below the Earth's surface.

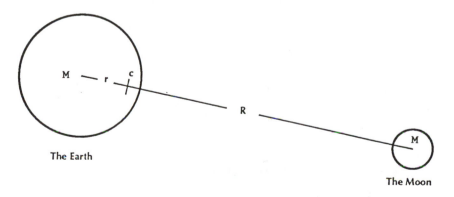

Fig. 5.5.6. The Earth-Moon Barycentre.

5.6 THE PRECESSION OF THE EARTH'S AXIS

In Section 1.6 the spinning of the Earth was likened to a spinning top which although it spins steadily, nevertheless under various external forces its axis wobbles or to use the technical term, precesses round a fixed point. This point in the case of the Earth's axis is called the North Pole of the Ecliptic and has the position in the celestial sphere of RA 18° (1^h 12^m) and Declination 66.5°. It is a help in understanding this, to mark this spot on your star globe or planisphere. The Earth's axis always points 23.5° away from this ecliptic pole and in the course of about 2,600 years it describes a full circle round it. During this century it points to within 1° of the star Polaris in Ursa Minor but about 4,000 years ago, the Earth's axis pointed to Alpha Draconis (see Fig. 5.6.1). This precession causes the point at which the Sun crosses the celestial equator to shift about 50 seconds of arc each year: a movement known as the Precession of the Equinoxes.

Draw on a star globe with a fine felt-tipped pen a circle of radius of 23.5° around the Pole of the Ecliptic to mark the path followed by the celestial pole. This circle represents the precession of the Earth's axis over a period of 26,000 years. The approximate dates can be marked to produce the figure shown in Fig. 5.6.1.

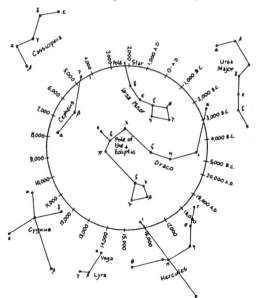

Fig. 5.6.1. The precession of the pole, showing the positions occupied by the pole at various dates between 5 000 BC and 20 000 AD. Reproduced from Sir James Jeans, The Stars in their Courses. By permission of Cambridge University Press, 1940.

5.7 THE EARTH'S AXIS AND THE PLANE OF THE ECLIPTIC

In section 2.17 the Sun and Moon 4000 years ago are noted as having bearings at rising and setting that are practically the same as they are today, whereas the precession of the Earth's axis causes Star positions to change considerably over the years, (see Fig. 5.6.1).

The explanation is that although the Earth's axis precesses round the Celestial Pole once in every 26,000 years and was well known to ancient sky watchers, the Earth's axis maintains an almost constant angle of 23.5° to the plane of the ecliptic which has changed by only half a degree over the past 4 millenia. This will explain how Stonehenge was used to study the Sun and the Moon, to make calendars, and predict eclipses, but shows no obvious evidence that the stones were used to study stars for any practical purposes, such as calendar making. Nevertheless, the very small change in the plane of the ecliptic of about 0.5° in 4000 years has been studied and used by astronomers and astrologers in helping us to date lunar and solar movements in various parts of the world. For those in doubt about this precession problem it is suggested that a better understanding can be achieved by a simple model or demonstration using a circle drawn on a table top to represent the Earth's orbit in the Plane of the Ecliptic, and a knitting needle stuck through an orange to represent the Earth's axis inclined at an angle of 23.5° to the vertical from the table, and going round the Sun in the centre of of the circle.

5.8 SPHERICAL TRIGONOMETRY USED IN ASTRONOMY

This section gives a convenient way to derive the basic formulae used in Astronomy and Navigation, and may be useful for a mathematics class as an introduction to vectors.

Star positions are defined by coordinates on the celestial sphere as shown in Section 3, using Altitude, Azimuth and Declination (δ). Hour angle and Latitude.

5.8.1 How these are related

The relations commonly used as calculator or computer programmes are:

$$\sin (\text{alt}) = \sin\phi \, \sin \delta + \cos \phi \, \cos \delta \, \cos \text{HA} \tag{1}$$

$$\sin \delta = \sin\phi \, \sin (\text{alt}) + \cos \phi \, \cos (\text{alt}) \, \cos \text{Az} \tag{2}$$

$$\tan (\text{Az}) = \frac{\sin \text{HA}}{\cos \phi \, \tan \delta - \sin\phi \, \cos \text{HA}} \tag{3}$$

In the usual spherical triangle used in astronomy, PZX of P, is the Celestial Pole, Z is the Zenith, and so denotes the Observer (who is apparently on top of the world), and X is the position of the star under observation.

The figure shows the parameters involved all of which are angles (Fig. 5.8.1). The Pole is at P the Azimuth at Z and the sides making angles at the centre of the sphere are respectively as follows:

ZP is the angle $(90 - \phi)$ ZX is the angle $(90 - \text{alt})$

PX is the angle $(90 - \delta)$

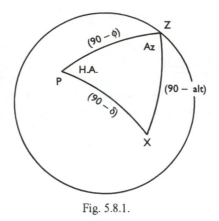

Fig. 5.8.1.

The relation (2) is the same form as (1) but with Az replacing HA and (90 – alt) replacing (90 – δ). Visualized by a simple clockwise rotation of the figure. The sides ZP, ZX and PX are all parts of great circles, and are expressed in degrees, but the HA at P is often expressed in Hours, minutes and Seconds (1 hour ≡ 15°).

In Fig. 5.8.2, which is a hemispere of *unit radius*, A, B, and C have the coordinates:

	x	y	z
(A)	0	0	1
(B)	sin c	0	cos c
(C)	sin b cos A	sin b Sin A	cos b

These are shown in the matrices that follow:

$$\overrightarrow{OB} = \begin{pmatrix} \sin c \\ 0 \\ \cos c \end{pmatrix}$$

and

$$\overrightarrow{OC} = \begin{pmatrix} \sin b \cos A \\ \sin b \sin A \\ \cos b \end{pmatrix} \quad \begin{matrix} \ldots x \\ \ldots y \\ \ldots z \end{matrix}$$

5.8.2 Scalar product

The scalar product of $\overrightarrow{OB}$ and $\overrightarrow{OC}$ can be expressed in two ways

$$\overrightarrow{OB} \cdot \overrightarrow{OC} = 1 \times 1 \times \cos a \tag{1}$$

and

$$\overrightarrow{OB} \cdot \overrightarrow{OC} = \sin c, \sin b \cos A + \cos c \cos b \tag{2}$$

so

$$\cos a = \cos c \cos b + \sin c \sin b \cos A \dots \tag{3}$$

Then using 5.8.1 and the Fig. 5.8.1, (3) becomes

$$\sin \delta = \sin \phi \sin \text{alt} + \cos \phi \cos \text{alt} \cos Az.$$

and by replacing Az by H.A. we have

$$\sin \text{alt} = \sin \phi \sin \delta + \cos \phi \cos \delta \cos \text{H.A.}$$

which are the basic relations for astronomical calclations relating declination, latitude, altitude, Hour Angle and Azimuth, and have been used to provide the computer printouts, graphs and nomograms of Chapter 3.

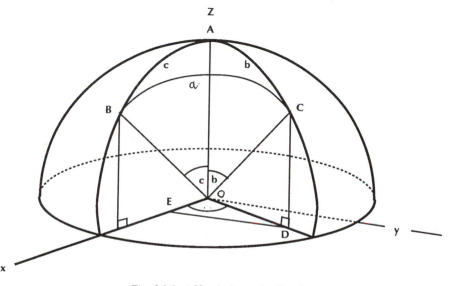

Fig. 5.8.2. A Hemisphere of unit radius.

5.8.3   Plane trigonometry

In plane trigonometry, the sides and angles of a triangle are related by the so-called Sine formula,

$$\frac{a}{\sin A} = \frac{b}{\sin B} = \frac{c}{\sin C}$$

There is a correspondingly elegant relation in spherical trigonometry which can be derived from Fig. 5.8.2 by considering the volume of the plane tetrahedron ABCO which is $\frac{1}{3}$ Area AOB × Perpendicular Dist of C from the plane AOB. This distance is DE which is the y ordinate of C and is from Fig. 5.8.2 = sin b sin a.

The volume of the tetraahedron is thus:

$$\tfrac{1}{3} \times \tfrac{1}{2} \times 1 \times \sin c \sin b \sin A = \tfrac{1}{6} \sin b \sin c \sin A$$

By applying this process to the sides AOC and BOC to derive the same volume, we find the volume of the same tetrahedron is $\frac{1}{6}$ sin c sin a sin B and $\frac{1}{6}$ sin a sin b sin C, respectively. We can equate these values. Then dividing each by sin a sin b sin c, we obtain

$$\frac{\sin A}{\sin a} = \frac{\sin B}{\sin b} = \frac{\sin C}{\sin c}$$

which is the Sine Rule for spherical trigonometry corresponding to the Sine Rule for plane trigonometry above.

Table 5.8.3. Useful Trigonometrical relations

(1)	sin (alt) sin φ sin δ + cos φ cos (HA)	Used to find altitude from φ, δ in HA OHA for alt φ δ.
	or, $\cos (HA) = \dfrac{\sin (alt) - \sin\phi \sin \delta}{\cos \phi \cos \delta}$	Follows from Pythagoras and plane trigonometry.
	When (HA) = 0 (or 180) we get, sin (alt) = sin δ sin φ + cos δ cos φ = cos (φ − δ) ∴ (alt) = 90 − φ + δ hence, meridian altitude	
(2)	φ = 90 − (alt) + δ	
	When (alt) = at rising and setting, then	
(3)	cos (HA) = − tan φ tan δ	This is the basic relation for astronavigation giving times of rising and setting.
(4)	sin δ = sin φ sin (alt) + cos φ cos (alt) cos (Az)	
(5)	or, $\cos (Az) = \dfrac{\sin \delta - \sin \phi \sin (alt)}{\cos \phi \cos (alt)}$	This is really of the same form as (1) with Fig. 5.8.2 rotated so that (alt) becomes δ and HA becomes (Az).
	When (alt) = 0 at rising and setting then sin (alt) = 0 and therefore	
(6)	$\cos (Az) = \dfrac{\sin \delta}{\cos \phi}$	Can be used for getting a bearing or checking a compass at sunrise or sunset.
(7)	$\sin (Az) = \dfrac{\sin (HA) \cos \delta}{\cos (alt)}$	This corresponds to the sine formula of plane trigonometry.
	used for finding a star's azimuth.	
(8)	$\cot (Az) = \dfrac{\cos \phi \tan \delta - \sin \phi \cos (HA)}{\sin (HA)}$	This is known as the four part formula, and can be deduced by the ordinary rules of plane trigonometry.

These relations are valid for both the northern and the southern hemispheres, provided values for φ in the southern hemisphere, and for δ when the celestial body is south of the celestial equator, are given correct negative values. Hour angles are always positive as they are measured from the meridian in the direction west, north, east, south. negative values for altitude may be encountered in some calculations, but this means that the value is below the horizon, and is measured towards the nadir.

5.8.4   Scientific calculations

Practically all problems in positional astronomy concerning hour angles (HA), latitude of the observer, ϕ, altitude (alt), declination δ, Right Ascension (RA), and azimuth (Az) can be solved by the scientific calculator using one or more of the relations of spherical trigonometry which are to be found in Table 5.8.3. (See conversion formulae in the *Handbook of the British Astronomical Association.*)

5.9   TO FIND THE LIMITING MAGNITUDE OF A TELESCOPE

The light gathering power of a telescope depends only on the diamter D of the objective. The night adapted eye pupil $\approx$ 8 mm in diameter.

The power of a telescope over the unaided eye is

$$= \frac{\text{Area of Objective}}{\text{Area of Eye Pupil}} = \frac{D^2}{8^2} \, .$$

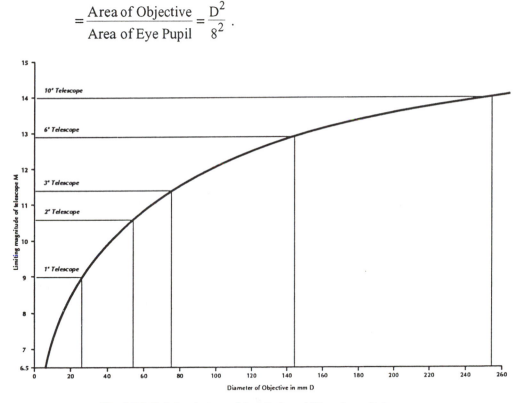

Fig. 5.9.1. Relation between Magnitude and Diameters of telescopes.

Our eyes perceive brightness in *logarithmic* steps, although physical optics deals in *arithmetic* steps.

We choose to recognize 5 steps between a bright star, A, (Magnitude 1) and a very faint star, B, (Magnitude 6). Physical optics shows in reality that star A is 100 times brighter than B. This can be determined in the following way.

If x is the interval the eye recognizes between A and B, then $x^5 = 100$. Taking logs,

$$5 \log x = \log 100 = 2$$

and

$$\log x = 0.4$$

our calculator gives

$$x = 2.512.$$

So a star of magnitude of say 4 is 2.512 times as bright as a star of magnitude 5.

What is the limiting magnitude (M) of your telescope? The limiting magnitude of the human eye is ≈ 6.5. The number of magnitude steps in illuminating power provided by a telescope will be $(M - 6.5)$, so by substitution we have,

$$2.512^{(M - 6.5)} = \frac{D^2}{8^2}.$$

Take logs

$$(M - 6.5) \log 2.512 = 2 \log D - 2 \log 8$$

$$(M - 6.5) \, 0.4 = 2 \log D - 2 \times 0.9$$

$$M - 6.5 = 5 \log D - 4.5$$

$$M = 2 + 5 \log D$$

So when D = 100, M = 12, and when D = 150 mm (6″), M = $2 + 5 \times 2.176 \approx 13$. Fig. 5.9.1 shows how M slowly increases with D (when D = 250, M = 14).

5.9.2   Magnitude of a stars

Norton's Star Atlas, 2000.0 has a Table (Number 40) listing the 26 nearest stars and another Table (Number 41) giving the 26 brightest stars. The positions and magnitudes of the stars are given together with their distances and spectral classes. The tables provide data that with enjoyable satisfaction can be checked using a calculator because the footnotes to the tables state that the parallaxes and the absolute magnitudes of many of the stars may not be in agreement with the parallax measurements, (see Appendix).

For example, the star Altair whose apparent magnitude (M_{app}) is 0.77, has an absolute magnitude (M_{ab}) 2.3: which is the magnitude a star would have if it were at a distance of 32.6 Light years or 10 parses. The actual distance of Altair is 16 Light years. So for the ratio

$$\frac{\text{apparent brightness}}{\text{absolute brightness}}.$$

We have accordingly

$$(2.512)^{(M_{ab} - M_{app})} = \left(\frac{32.6}{16}\right).$$

The left hand side is the ratio calculated by the formula for brightness, the right hand side is simply the Inverse Square Law governing light intensity.
The Left hand side

$$= (2.512)^{(2.3 - 0.77)}$$
$$= (2.512)^{1.52}$$
$$= 4.092$$

The Right hand side

$$= 4.15$$

A few examples worked in this way will help students to understand and make use of the terms used in Altases under Magnitude, distances luminosity and parallax.

5.10   THE DIP OF THE HORIZON AT SEA AND THE RADIUS OF THE EARTH

The Earth's spherical shape has been studied and used for centuries by seamen and navigators to measure distances across the sea.

Lighthouses are shown on all navigational charts with their heights above sea level in metres. A person on a small boat at B approaching the coast suddenly sees the light from the lighthouse LX flashing on the horizon. A simple piece of trigonometry will give the distance of the boat from the shore, BX, provided the radius of the Earth is known (= 6378 km).
The angle θ

$$= arc \cos \frac{BC}{CL}$$

but

$$Cl = CX + LX$$

so

$$\cos \theta = \frac{6378}{(6378 + 0.02)}.$$

Your calculator will give $\theta = 0.143485°$. This angle expressed in minutes of arc is 8.61' and each minute of arc on the Earth's surface is equal to 1 Nautical Mile 1.852 km. Thus, the distance BX = 15.95 km.

This calculation is an example of the great value of the calculator, as formerly this calculation would have required the laborious use of 6 figure log and trig tables.

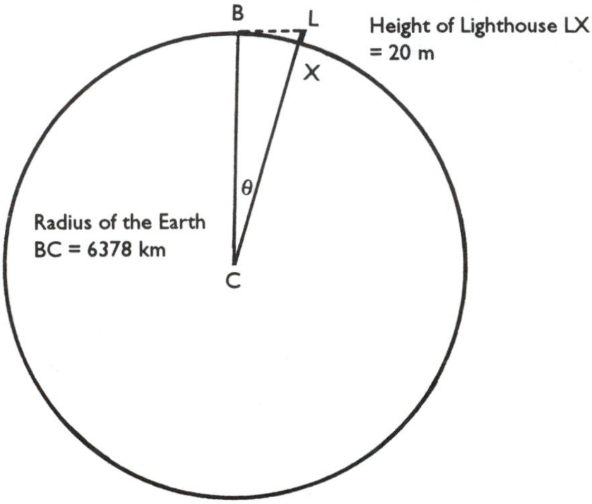

Fig. 5.10. The dip of the horizon at sea level.

5.11 A DEEP BREATH

To bring into this calculation section a touch of history, physical chemistry, Earth science, as well as mind boggling figures, consider 1 litre of air, for example, Julius Caesar's last breath.

This contained 3×10^{22} molecules. Assume that these molecules after 2000 years in a turbulent atmosphere are thoroughly mixed with the rest of the atmosphere. The number of litres of air in the Earth's atmosphere can be roughly calculated by multiplying the surface area of the Earth by the mean thickness of the atmosphere which becomes very thin after about 5 kilometres. Taking 5 km as the mean thickness, the number of cubic metres of air in the atmosphere (Fig. 5.1)

$$= 4 \, (6.368 \times 10^6)^2 \times 5 \times 10^3$$
$$= 2.55 \times 10^{18} \, \text{m}^3.$$

There are 10^3 litres in one cubic metre, so the number of litres of air in the atmosphere is 2.55×10^{21}. We conclude by this rough calculation that there are about ten times more molecules in one litre of air than there are litres of air in the atmosphere. So assuming perfect mixing, every time you take a deep breath, you probably inhale about ten molecules of Julius Caesar's last breath, and incidentally of any of his contemporaries, such as Anthony and Cleopatra!

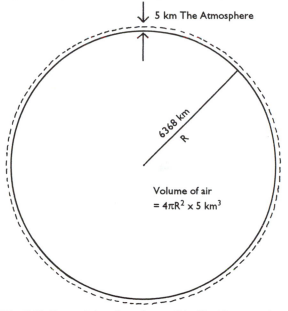

Fig. 5.11. Determining the volume of the Earth's atmosphere.

5.12   SPHERICAL AND PARABOLIC MIRRORS

This section may provide an instructive exercise in physics and trigonometry involving spherical and parabolic mirrors of telescopes, and a calculator. The four diagrams, Figs 5.12.1, 5.12.2, 5.12.3 and 5.12.4, together with with Table 5.12.1 have been devised to help teachers, students and potential mirror grinders deal with these questions quantitatively to which often vague qualitative answers are given.

For those who might wish to grind a mirror objective for a telescope a little geometry will give a rough idea of the amount of grinding, d, to be done on a piece of plane glass EDB to form a spherical mirror EAB which has its centre at C, a diameter EB (radius r), and a radius of curvature 2f.

By Pythogoras' theorem,

$$EC^2 = ED^2 + DC^2$$

where

$$EC = 2f$$

so

$$4f^2 = r^2 + (2f - d)^2$$

$$4df = r^2$$

neglecting d^2 which is small. If the focal length of the mirror is to be 1000 mm and $r = 75$ mm. Then

$$d = \frac{r^2}{4f} = \frac{5625}{4000} = 1.40625 \text{ mm}$$

The distance d can be checked by placing a straight edge BE across the mirror and using a feeler gauge. It will be appreciated that grinding a glass surface to this depth and figure using grinding powders requires great patience and skill, but for a high quality mirror the curved surface must be ground a little more to produce a paraboloid surface and this will provide an instructive exercise in physics and trigonometry to determine the precise differences between a spherical and a parabolic mirror

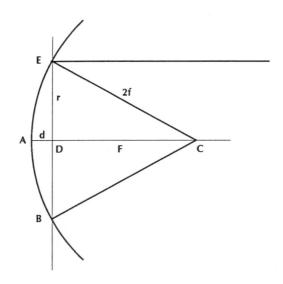

Fig. 5.12.1. Grinding a spherical mirror objective.

5.12.2 Spherical aberration of spherical mirrors

A spherical mirror is known to have serious limitations because of its spherical aberration that is responsible for the focus of the mirror being spread out along the axis towards the pole A. How much is this spread and does it depend on the focal ratio of the mirror? (See column 2 of Table 5.12.5.)

In Fig. 15.12.2, XE is a ray parallel to the axis and is incident on the mirror AEG at E at an angle ϕ. EC is a radius of the circle AEG with its centre at C. The angles ϕ are marked. EQC is isoceles and EQ = QC, EQ is the reflected ray.

$$EQ = QS + f$$
$$ED = EQ \sin 2\phi$$
$$\quad = 2f \sin \phi \, .$$
$$QS = \frac{2 f \sin \phi}{\sin 2 \phi} - f$$

The longitudinal spread of the focus, QS

$$= f \left(\frac{2 \sin \phi}{\sin 2 \phi} - 1 \right)$$

as in column 2 of Table 5.12.1. This spherical aberration is overcome in a parabolic reflector as shown in Fig. 5.12.3.

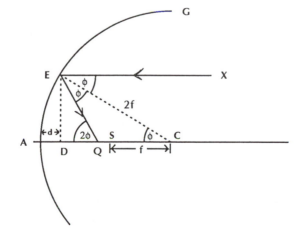

Fig. 5.12.2. The spherical aberration of a spherical mirror.

5.12.3   The paraboloid

A proof that a paraboloid will focus all rays that are parallel to the axis to a single point S. It is generally accepted, but rarely proved, that a parabolic mirror overcomes this spherical aberration of a spherical mirror, by bringing to a single focus all the rays entering the mirror parallel to its axis. How can this facility be explained using the mathematics of the parabola and the laws of reflection of light?

The parabolic mirror in the figure has the shape to bring all rays incident on the mirror parallel to the axis AC to a single point S, the geometrical focus of the parabola APV.

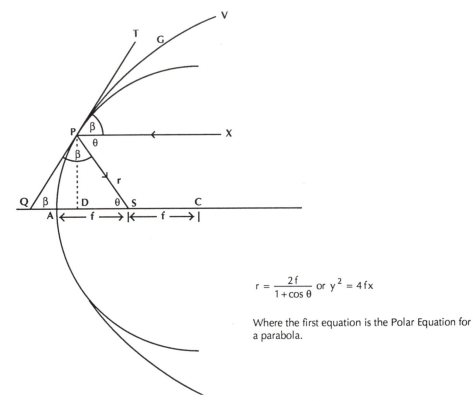

$$r = \frac{2f}{1 + \cos \theta} \text{ or } y^2 = 4fx$$

Where the first equation is the Polar Equation for a parabola.

Fig. 5.12.3. A proof that a paraboloid will focus all rays that are parallel to the axis to a single point S.

Consider the ray XP incident on the mirror at P. QPT is the tangent to the parabola at P and has a slope β = XPT and = SPQ by the reflection law. From the figure,

$$2\beta = 180 - \theta$$

or

$$\beta = 90 - \tfrac{1}{2}\theta .$$ (1)

The angle β has a tangent given by dy/dx which for the parabola $y^2 = 4fx$, $\tan \beta = 2f/y$.

The radius vector of the parabola

$$PS = \frac{2f}{1 + \cos \theta} \quad \text{as in column 5 of Table 5.12.5.}$$

The polar equation for a parabola

$$y = PD = PS \sin \theta .$$

So,

$$\tan\beta = \frac{2f}{PS \sin \theta}$$

$$= \frac{(1 + \cos \theta)}{\sin \theta} .$$

This is a trigonometric identity, with the solution $= 90 - \tfrac{1}{2}\theta$ as in (1) above.

So the optics of the reflector and the geometry of the parabola combine to bring all parallel rays to a common focus at S.

5.12.4 A formula for the gap between a spherical mirror and a corresponding parabolic mirror as a function of the focal ratio

Fig. 5.12.4 shows a circle AEH of radius 2f, and a parabola APV having the polar equation

$$r = \frac{2f}{1 + \cos \theta}$$

with its focus at S and having contact with the circle at A. They maintain a close contact so long as θ is small: less than 3°. At A the circle and the parabola have the same radius of curvature. At focal ratios less than about 6 the curves rapidly separate as shown in column 7 of Table 5.12.1. From Fig. 5.12.3 the gap between the circle and the parabola. PE = PS − ES where

$$PS = \frac{2f}{1 + \cos \theta}$$

which is the Radius Vector of a Parabola

$$ES \sin \theta = EV = 2f \sin \phi$$

$$ES = \frac{2f \sin \phi}{\sin \theta} \quad \text{(column 6 of Table 5.12.5)}$$

$$PE = 2f \left(\frac{1}{1 + \cos \theta} - \frac{\sin \phi}{\sin \theta} \right) \quad \text{(column 7 of Table 5.12.5)}$$

The corresponding focal ratios

$$\frac{f}{\text{Diameter}} = \frac{1 + \cos \theta}{4 \sin \theta}$$

as shown in column four of Table 5.12.5.

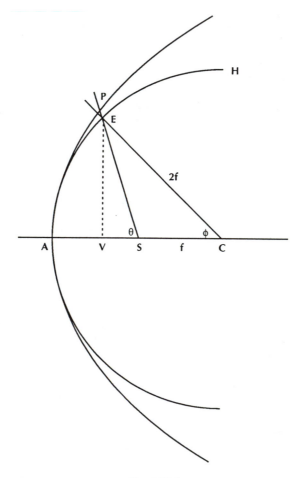

Fig. 5.12.4.

Table 5.12.5

1	2	3	4	5	6	7
ϕ	Spread of focus QS mm	θ	f/Diameter	Rad. Vector of parabola PS	ES	PS–ES
In degs.	$f\left(\dfrac{2\sin\phi}{\sin2\phi}-1\right)$	Arc tan $\dfrac{2\sin\phi}{2\cos\phi-1}$	$\dfrac{1+\cos\theta}{4\sin\theta}$	$\dfrac{2f}{1+\cos\theta}$	$\dfrac{2f\sin\phi}{\sin\theta}$	Approx. gap between OC
1	0.1523	1.99969	14.32	1000.3045	1000.3043	<0.00001
2	0.6095	3.99756	7.17	1000.2179	1000.2176	0.0003
5	3.8198	9.96228	2.86	1007.5963	1007.5818	0.0145
10	15.4267	19.70648	1.44	1030.1672	1029.3630	0.8042
15	35.2762	29.05191	0.96	1067.1333	1065.9721	1.1612
20	64.1838	34.87799	0.73	1117.7430	1114.1043	3.6382
30	154.7005	53.79398	0.49	1257.3156	1239.3136	18.0020
45	414.2135	73.67505	0.33	1561.1770	1473.6257	87.5513
60	1000.000	90.0000	0.25	2000.0000	1732.0580	267.942
2.15	**0.7387**	**4.29518**	**6.68**	**1001.4063**	**1001.4057**	**0.0006**

The **Bold** line applies to reflector 150 mm diameter, focal length 1000 mm.

5.13   AN ANALYSIS OF THE EQUATIONS OF TIME

A method for forming the equations for plotting the graphs of Fig. 2.10.1, showing how the values of E_1 and E_2 (y axis) vary with the date throughout the year, (x axis), is given below.

There are two phenomena mainly responsible for the equation of time E.

(1)   The Earth moves round the Sun in an elliptical orbit. This component of E (Fig. 5.13.1) is ascribed to the eccentricity of the Earth's orbit, and is referred to as E_1.

(2)   The plane of the Earth's orbit is inclined to the plane of the Earth's Equator. This is ascribed to the obliquity of the ecliptic, E_2, as shown in Fig. 5.13.2.

5.13.1   The effect of the eccentricity of the Sun's apparent orbit
The eccentricity e of the ellipse is defined as

$$CE/CA = e = \tfrac{1}{60} \text{ approx.}$$

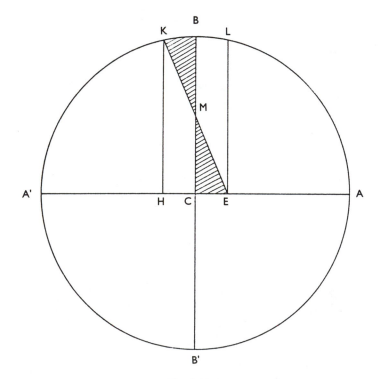

Fig. 5.13.1.

The mean Sun that gives us our time, apparently travels uniformly round a path in the plane of the Earth's equator and makes equal angles in equal times. At L the Sun appears to have gone through 90°, see Fig. 15.13.1.

The apparent true Sun, however, has travelled faster because by Kepler's 2nd Law it describes equal *areas* in equal times.

So while the mean Sun goes from A to L, the true sun goes from A to K, also,

Area KMB = Area CEM.

The true Sun describes the area AKEA in a $\frac{1}{4}$ of the periodic time and AKEA is $\frac{1}{4}$ the area enclosed by the orbit.

ABCA is also $\frac{1}{4}$ the area of the orbit since area KMB is equal to area CEM.

The true Sun has moved through an angle KEL more than the mean sun measured in the ecliptic plane

tan KEL = 2/60 and KEL = 1°.91.

This is the difference in longitude between true Sun and mean Sun, which is approximately equal to the difference in Right Ascension at this point in the orbit. 1° in (RA) = 4 min, so in the situation shown in Fig. 5.13.1, the RA true Sun in 1.91 × 4 m, about 7.64 m ahead of the mean Sun, or 7 m 38.4 s.

It can be seen from Fig. 5.13.1 that as the Sun apparently moves fastest round the Earth when it is nearest the Sun (at perigee A), the Sun will appear to gain on the apparent mean Sun: which by definition moves at a uniform *angular* velocity in the plane of the equator. The true Sun and the mean Sun, however, appear level again at A', and this means that the true Sun must move more slowly in its orbit after reaching the half-way mark, i.e., when the angle LEA is 90°. The difference between 'true sun time' and the 'mean dynamical sun' by Fig. 5.13.1 is a maximum at 90°.

It is reasonable to assume that the maximum difference in position, and therefore in time, will occur when AEL = 90°. This suggests that the value of this difference in position varies sinusoidally from 0° to 180° from *2nd January to 2nd July (perigee to apogee)*.

The angle to be used in the sine relation for any date is ω t where ω is the mean angular velocity (degrees per day) of the Earth in its orbit, and t is the number of days that have elapsed from about 2nd January to the date for which the advance or retardation in time is required. We can make approximate calculations using this simplified principle.

The advance in HA is therefore −7.64 sin ω t (from above). ω for practical purposes is 360 in 365.25 days = 360/365.25. As used in Fig. 2.10.1.

The expression shows that on 2nd January $E_1 = 0$

When	$\omega t = 90$	about 1st April	$E_1 = -7^m.64$ or $-7^m38^s.4$
	$\omega t = 180$	about 2nd July	$E_1 = 0$
	$\omega t = 270$	and on 1st October	$E_1 = +7^m.64$ or $+7^m38^s.4$

The effect of the eccentricity of the Earth's orbit on the Sun's time keeping E_1) can best be shown on a graph for $E_1 = -7^m.64 \sin \omega t$ as shown in Fig. 2.10.1.

5.13.2   The component of E caused by the obliquity of the ecliptic, E_2

Quite separate from the irregularity of the Sun's time-keeping due to the ellipticity of the Earth's orbit round the Sun, is the irregularity caused by the fact that the Sun appears to move during the course of a year in a path on the ecliptic which is inclined to the equator at an angle of $23°.44$, which has the symbol ε. The equator is really the basis of our time measurements. Our daily time system is based on the Earth's revolution round the polar axis, and the equator is the plane at right angles to this axis.

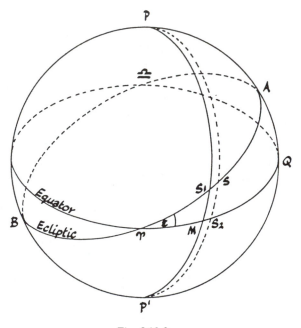

Fig. 5.13.2.

It can be seen from Fig. 5.13.2 that a about 21st March ♈ and 23rd September, ♎, the Sun, during one day of progress along the ecliptic, will travel $1°$, but along the equator this will mark out only $1° \cos \varepsilon$ (where $\varepsilon = 23°.44$), i.e., $1° \cos 23°.44 = 0°.917\ 477$.

The Sun during this day will thus lose $1°.0 - 0°.917\ 477 = 0°.082\ 58$ of RA but will gain $0°.082\ 58$ of HA because RA and HA are measured in opposite directions.

In Fig. 5.13.2 S is the true sun, S_1 is the dynamical mean Sun which appears to move uniformly in its annual motion in the plane of the ecliptic, and S_2 is a point on the equator so that $S_2 =$ S_1.

S_2 is the *astronomical mean Sun* with apparent motion in the equatorial plane. E_2 is the part of the equation of time due to the obliquity of the ecliptic. E_2 = the difference in hour angle between S_1 and S_2 i.e., angle S_1PS.

This can be regarded as the advance of the 'ecliptic mean Sun' over the 'equator mean Sun', or the advance of the 'dynamical mean Sun' over the 'astronomical mean Sun'. The dynamical mean Sun in the ecliptic, and the astronomical mean Sun in the plane of the equator are *in step* when $\Upsilon S_1 = \Upsilon M$, i.e. at longitudes $0°$, $90°$, $180°$ and $270°$.

We have so far merely made a rough qualitative estimate of the effect of the obliquity of the ecliptic (E_2) on the equation of time.

We treated a one day apparent movement of the Sun from Υ as making a small plane triangle with the equator; while this is not greatly in error for a $1°$ movement from Υ, it will be badly in error for a time of one month. We must regard $\Upsilon S_1 M$ as a *spherical* triangle (Fig. 5.13.2).

If we apply the four part relation Table 5.8.3 (8) then,

$$\cos \Upsilon M \cos \varepsilon = \sin \Upsilon M \cot \Upsilon S_1 = \sin \varepsilon \cot S_1 M \Upsilon$$

where

$$\varepsilon = 23°.44$$

but

$$S_1 M \Upsilon = 90°$$

and

$$\cot S_1 M \Upsilon = 0,$$

therefore

$$\tan \Upsilon S_1 \cos \varepsilon = \tan \Upsilon M$$

$\cos \varepsilon$ is a constant and $= 0°.917\ 47$.

The dynamical mean Sun (in the ecliptic) and the astronomical mean Sun (in the equator) are in step when $\Upsilon S_1 = \Upsilon M$, i.e., when they both equal $0°$, $90°$, $180°$ and $270°$. there would appear to be a maximum difference between the two when $\Upsilon S_1 \approx 45° =$ longitude of the mean dynamical Sun, and this can be checked using a calculator. From relation

$$\tan \Upsilon M = 0°.917\ 47 \tan 45°$$

$$\Upsilon M = 42°.536$$

then

$$E_2 = \Upsilon S_1 - \Upsilon M = 45° - 42°.536 = 2°.464\ (1 = 4 \text{ minutes})$$

$$= 9.856 \text{ minutes.}$$

We estimated the maximum value might occur when the longitude of the mean Sun was $45°$. This is very near the longitude for maximum E_2, but it can be shown by trying longitude $46°$ and then $47°$ that E_2 (max) occurs at longitude $46°$. There are doubtless more rigorous ways of finding the Sun's longitude for E_2 (max) but this simple calculator check is adequate for our purpose.

As E_2 varies approximately sinusoidally with $2(RA)$, we may take the value of E_2 to be given by

$$E_2 = E_2 \text{ max } \sin 2(RA)$$

i.e.,

$$E_2 = 9.863 \sin 2(RA)$$

The factor of 2 in $2(RA)$ is required as $E_2 = 0$ when $(RA) = 0°, 90°, 180°$ or $270°$ and is a maximum when $\sin 2(RA) = 1°$, i.e., when $(RA) = 45°$ or $135°$.

E_2 changes sign four times in the course of one year and E_1 changes sign only twice in a year, as is shown in Fig. 2.10.1.

5.13.3

We now have a workable approximate formula for finding E the equation of time from the date and the Sun's RA on that date, using relations [8.7(2)] and [8.7(3)]

$$E = E_1 + E_2$$

$$= -7.64 \sin \gamma \, t + 9.863 \sin 2 \, (RA). \qquad\qquad [8.8(1)]$$

It is a rewarding and instructive exercise for the calculator to plot three graphs from the results obtained from the separate components of Fig. 5.13.3.

$$E_1 = -7.64 \sin \gamma \, t$$

$$= -7.64 \sin \left(\frac{360}{365.25} \times \text{number of days since 2nd January} \right)$$

The graph E_1 is shown in Fig. 2.10.1.

Plotting E_2 from $E_2 = 9.864 \sin (2 \times RA))$ is straightforward and the graph is also shown in Fig. 2.10.1.

E is simply $E_1 + E_2$ and shows the familiar graph for the equation of time, as shown in Fig. 2.10.1.

The values for E can be checked from the *Nautical Almanac* from *Whitaker's Almanac* or from the *Handbook of the British Astronomical Association,* (explanation supplement).

As an example, the calculation for finding the equation of time for October 31st is shown.

The number of days from 2nd January to 31st October are 302.

This is reduced to degrees by $302 \times 360/ 365.26 = 297.65$.

The RA on this day from *Whitaker's Almanac* is shown as $14^h 21^m = 215.25$ and $2 \times RA = 430.5$ so $E_1 + E_2 = 7.74 \sin 297.65 + 9.864 \sin 430.5 = 6.7675 + 9.2982 = 16.065$ minutes, the accepted value.

6

Appendices and useful information

6.1   ADVICE FOR BEGINNERS

Among the many activities for sky watchers to do, that would prove to be perhaps, the most helpful and rewarding of all is to join an astronomical club or society. Many towns and cities have a local society at which you can meet other people who are interested and who would be pleased to share with you their sky watching experiences, their telescopes, star charts, plani-spheres or almanacs. See Section 6.8 under the heading *Suggestions for Further Reading*, which mentions the *Handbook for Astronomical Societies*; this gives the addresses of about 200 U.K. astronomical societies, and includes sources of information and equipment.

Many of these societies have a wide range of membership from those with professional experience and qualifications to those who might have difficulty in finding Polaris on a clear night! So by joining you will not only have opportunities to learn, but also be able to help others. Observing can have its pleasures and rewards enhanced when undertaken in small groups.

6.2   GIVING A TALK

Members of astronomical societies or enthusiastic hobby sky watchers are often called upon to give a talk. This can be a helpful and satisfying thing to do. Here are some tips on basic principles to follow for a successful talk.

(1)   **Know your subject**, but while expert knowledge of this is *necessary* for a good talk, it is not *sufficient*. It alo requires skills in communication.
(2)   **Have a well-defined and limited objective**, which should be stated at the beginning. Don't spend valuable time apologizing for your inadequacy for the task before you, or explaining at length how difficult it is to deal with your subject in the short time at your disposal. The audience will not be interested in this, so get on with achieving your objective.

(3)   **Prepare your talk** to suit the level and interests of the audience. Have a good opening paragraph, and arrange your material into a logical sequence. It is useful to end with a summary of the main points, or the message you have to deliver.

(4)   **Prepare your notes** to remind you of the key points so that you can refer to them easily without papers getting mixed up or falling on the floor. To write out your talk fully may be useful, but *don't read it* or it will sound rather dull. Try out your talk on a friend for timing and then trim for length, leaving a few minutes for questions or discussion.

(5)   **Use equipment, models, slides, charts and the chalk-board whenever possible**. "A picture is worth a thousand words" or, as in boxing, "one in the eye is worth two on the ear".

(6)   **Whatever you show must be clearly seen by all**. The limit of resolution of the eye is 2 minutes of arc, but audiences are often treated to detailed information in letters and numerals on screens or charts that are well below this limit. Lettering, however it is presented to viewers at around 15 metres distance, should be at least 30 mm high. Users of the overhead projector are tempted to project transparent photocopies of tables and graphs in small print from books, so that lettering appears on the screen less than 5 mm in height. This is a source of irritation to your audience, as well as a dead loss to them.

(7)   **Don't use the blackboard or an overhead projector transparency as a scribbling pad**. Prepare diagrams beforehand. Always use a pointer on a chalkboard or on a screen projection, and stand clear. Use a knitting needle to point to detail on a transparency; your finger looks grotesque and is distracting. Avoid, at all costs, standing in the dark and pointing with a finger to a feature on the screen. You then become the only person in the room who knows what you are talking about!

(8)   **Don't talk to the blackboard**, or to the top of the table. You should look at your audience. Your notes should be on a stand at chest height, with a small lamp that lights up your notes and your face even when slides are being shown. You are, yourself, an important audio-visual aid.

(9)   **Avoid distracting mannerisms**, such as jingling coins in your pocket or putting on the caged-lion act.

(10) **Stand up, head up, and speak up loudly and clearly enough to be heard comfortably at the back row**. Don't shout out, *"Can you hear me at the back"*, but make sure you can be heard by a little practice. If you have to use a microphone, also make sure that you understand its use before you start and that it is working properly.

(11) **If stuck for a word or phrase, pause deliberately**. This is less annoying and more effective than filling in gaps with dummy words such as 'umm' or 'err'. This can become a habit. Don't mumble or let your voice drop at the end of each sentence.

(12) Astronomy is an international science, so **use only the internationally accepted units known as SI units**, as used for example, in *The Handbook of*

the British Astronomical Association. Wavelengths of light-waves are now expressed in nanometers (10^{-9} m).

(13) **Slides have a habit of appearing on the screen upside down or inside out.** A slide can be put into the carousel in eight different ways, of which only one is correct. To ensure that the slide goes in properly, hold it up to the light so that it is seen as you want it to appear on the screen. Make a clear mark at the bottom left-hand corner with a felt-tipped pen as shown in Fig. 6.2.1.

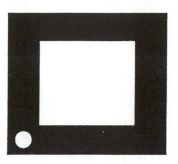

Fig. 6.2.1. First mark your slide.

Then rotate the slide through 180° in the plane of the slide so that the ink mark appears at the top right-hand corner of the slide as shown in Fig. 6.2.2, which is the correct position for placing in the carousel ready for projection.

Fig. 6.2.2. Rotate the slide through 180° and then insert in the carousel.

(14) **Finally**, and in continuation of tip number 10 above, we might add, with no intended discourtesy, "Shut up". If you are asked to speak for, say, 20 minutes, or for 40, finish five minutes before time to allow for questions or discussion which are important parts of your talk and should not be squeezed out. If you are one of several speakers, to encroach on the time allotted to your successor is inconsiderate.

6.3 GLOSSARY

All terms which appear in *italics* are included in this glossary.

Absolute magnitude: the magnitude which a star would have if it were at a distance of 32.6 *light-years* or 10 parsecs from us. It is a measure of the real brightness of a *star.*

Altazimuth: the name given to a telescope mounting which allows a telescope to move up and down in *altitude* and horizontally in *azimuth*.

Altitude: the vertical distance of a celestial body measured vertically upwards from the horizon.

Aphelion: the most distant point from the Sun reached by a *planet* on an elliptical *orbit* in the Solar System.

Apparent magnitude: the apparent brightness of a celestial body as it appears to an observer.

Astronomical Unit; AU: The mean distance between the Earth and the Sun.

Azimuth: the angle of a celestial body measured clockwise from the north point round the horizon.

Binary system: a pair of stars orbiting around one another.

Black hole: an extremely superdense celestial body emitting no *radiation*.

Celestial equator: the equator of the *celestial sphere*. It is a *great circle* and is the projection on the sphere of the Earth's equator.

Celestial sphere: an imaginary sphere centred on the Earth and with the stars on its inside surface. It is used for defining star positions.

Comet: a collection of ice and rocklike particles which orbits the Sun in an elliptical path. Comets may display tails when close to the Sun.

Conjunction: the apparent close approach in the sky of celestial bodies.

Constellation: A group of stars which makes a recognised pattern in the sky.

Culmination: the passage of a star across an observer's celestial meridian also known as the transit of meridian.

Declination: often marked by the symbol δ (delta) is the angular distance of a celestial body north or south of the *celestial equator*. It is analagous to latitude on an Earth globe.

Eclipse: a total or partical disappearance of a celestial body in the shadow of another: Initially an eclipse of the Sun by the Moon.

Ecliptic: the apparent path of the Sun on the *celestial sphere*.

Ellipse: a closed figure which curves around two points (foci). It can be formed by slicing obliquely through a cone. Planetary *orbits* are all ellipses.

Elongation: the angle between the Sun and a celestial body.

Equatorial mounting: a telescope mounting in which the telescope rotates about one axis parallel to the Earth's polar axis and also rotates at right- angles to it (i.e., in *declination*). It allows the telescope to track the apparent motion of a celestial body in a single movement about its polar axis.

Equinox: when day and night are equal (i.e., 12 hours each). At these points in the year, the Sun is at those places where the *celestial equator* and the *ecliptic* cross one another.

Galaxy: A remote island of millions of stars.

Gamma (γ) rays: very short wave radiation, more penetrating than *X-rays*. (Wavelength 10^{-14} to 10^{-12} metres.)

Gravity: the force with which all bodies attract one another.

Great circles: a circle on a sphere with a diameter equal to that of the sphere. The *celestial equator* is a great circle on the *celestial sphere* and the terrestrial equator is a great circle on the Earth.

Infra-red radiation: *radiation* with a wavelength longer than red light (10^{-6} to 10^{-4} metres).

Knot: an international unit of speed used in navigation. It is 1.852 km per hour. 1.852 km is 1′ of arc on the Earth's surface, known as 1 nautical mile.

Light-year: the distance light travels in one year: 9.4607×10^{12} km.

Magnetic field: the region of magnetic influence of a magnet or of a celestial body which acts like a magnet. An electric current has an associated magnetic field.

Magnetosphere: the *magnetic field* around a celestial body.

Meteor: a small piece of rock or metal which comes from outer space and falls towards the Earth's surface. Friction of the air causes it to heat up and emit light. A meteor ordinarily burns away before it reaches the Earth's surface.

Meteorite: material which has not been consumed comletely in its passage through the atmosphere and which therefore lands on the ground.

Nadir: the point in the celestial sphere which is diametrically opposite the observer's Zenith.

Neutron star: a very dense compact star which often behaves like a *pulsar*.

Nomogram: a system of graphs showing relations between 3 or more variables.

Nova: a star which throws off a shell of hot gas and thereby appears brighter for a short time, looking as if a new star had been born.

Occultation: an occultation occurs when one celestial body moves in front of another.

Opposition: when a *planet* lies on the side of the Earth opposite to the Sun.

Orbit: a path of one body moving around another.

Parabola: a curve which is closed at one end and open at the other. Such a curve can be obtained by cutting a slice down a cone parallel to one side.

Parallax: the apparent shift in position of a body when it is observed from two different positions.

Parsec: a distance of 3.26 *light-years*, i.e., the distance at which a star would display a *parallax* of 1 arc second.

Perihelion: the nearest point to the Sun on an elliptical *orbit* of a *planet* in the Solar System.

Planet: in astronomy the term is usually used to describe a non-radiating body in *orbit* around a star.

Precession: the movement of the point of intersection of the *celestial equator* and the *ecliptic*. This effect is caused by the movement of the Earth's polar axis.

Prominence: a flame-like projection of hot, glowing hydrogen gas sometimes seen at the Sun's limb during a total solar eclipse and, at other times by using special equipment such as a spectrohelio-scope.

Pulsar: an extremely dense collapsed star which emits rapid pulses of *radio waves* and sometimes other *radiation.*

Quasar: a 'quasi-stellar' source, i.e., a body which looks like a star but which emits far too much *radiation* to be one. It may be a bright core of a *galaxy.*

Radiation: a term for 'electromagnetic radiation', which ranges from *radio waves* to *gamma rays* and includes light.

Radio waves: long wavelength *radiation*; in astronomy the range goes from a few millimetres up to 30 metres.

Red giant: a very large star which radiates mainly red light.

Refraction: the bending of a light ray when it passes from one transparent substance to a denser or less dense transparent substance.

Right Ascension: angular distance of celestial objects measured eastwards along the *celestial equator* from the vernal equinox g .

Satellite: a natural or artificial body in *orbit* around a *planet.*

Sidereal Time: time measured by the Earth's rotation with reference to the stars instead of to the Sun. A sidereal day is about 4 minutes shorter than a solar day.

Solar wind: an outflow of atomic particles from the Sun, travelling through the Solar System.

Solstice: a 'standstill point' applied to the Sun when it changes the direction of its apparent movement in *declination*. At the Summer solstice the Sun has its maximum declination ($\delta = +23.5$), at the Winter solstice its declination is at its minimum ($\delta = -23.5$).

S.I. Units: the International System of units used for all scientific work, and using the basic units — metre, kilogram and second.

Spectroscope: an instrument for looking at *spectra.*

Spectrum: the range of wavelengths of a particular section of electromagnetic radiation. The spectrum of sunlight runs from red (long wavelength) to violet (short wavelength).

Star cluster: a group of stars held together in a cluster together by the gravitational forces between them.

Synodic period: the period between two successive alignments of celestial bodies. For a *planet* the synodic period is the time between successive *conjunctions* with a particular star or between *oppositions.*

Transit: a passage of a celestial body across the disc of another (e.g. the transit of Venus across the face of the Sun), or its passage across a given point (e.g. the transit of a star across the meridian or south point); when a star crosses the meridian it is said to transit the meridian.

Ultraviolet light: light shorter in wavelength than violet light but longer than *X-rays.* (Range about 10^{-6} to 10^{-8} metres.)
Universal Time: The mean solar time on the Greenwich meridian.

Variable stars: stars which vary in brightness over specific periods of time.

White dwarf: a very compact star close to the end of its life and which emits white light. A white dwarf is not as dense as either a *pulsar* or a *black hole.*

X-rays: short wave *radiation.* (Range about 10^{-8} to 10^{-11} metres.)

Zenith: the point directly overhead.

6.4   A NOTE ON NOMOGRAMS

This note supplements the section on Nomograms to be found in Chapter 3 and is reproduced here to prevent the reader having to turn back to the pages in question.

(1)   **The alignment nomogram**
   An ordinary graph shows how two variables can be related. Relations having three variables can be represented by three calibrated parallel straight lines, each representing one of the variables. In Fig. 3.7.1, a straight line drawn across the diagram gives the value of any one parameter in terms of the other two.

(2)   **The intersection nomogram**
   In order to depict the mathematical relations between more than three variables, an intersection nomogram can be used. Fig. 3.7.2 shows this type of nomogram (for a constant latitude of 51°N) consisting of two sets of curves, one set for altitudes, the other for Azimuths, and also two sets of straight lines depicting Declinations along the y axis and Local Hour Angle along the x axis, and giving two intersection points, P and Q, as examples.

   Tables of coordinates for plotting the intersection nomograms for Latitude 40°N, suitable for either Southern Europe or the Mid States of the USA, and intersection nomograms for Latitude 57°N, suitable for Scotland, are given in Tables 6.4.1, 6.4.2, 6.4.3, 6.4.4.

Table 6.4.1. Table of co-ordinates of points for plotting altitude cruves, or Almucantars for latitude 40° N†.

Declination (δ)		Altitude																
		5	10	15	20	25	30	35	40	45	50	55	60	65	70	75	80	85
	80	00.00	00.00	00.00	00.00	00.00	179.98	116.54	85.79	56.16	00.02	00.00	00.00	00.00	00.00	00.00	00.00	00.00
	75	00.00	00.00	00.00	00.00	179.99	127.57	103.80	83.66	64.22	42.93	00.02	00.00	00.00	00.00	00.00	00.00	00.00
	70	00.00	00.00	00.00	179.99	133.82	113.39	96.67	81.49	66.83	51.80	34.81	00.02	00.00	00.00	00.00	00.00	00.00
	65	00.00	00.00	179.99	137.99	119.61	104.78	91.59	79.28	67.38	55.48	43.05	28.89	00.01	00.00	00.00	00.00	00.00
	60	00.00	179.99	141.04	124.08	110.49	98.51	87.47	77.01	66.87	56.86	46.74	36.13	24.10	00.00	00.00	00.00	00.00
	55	179.99	143.43	127.54	114.83	103.68	93.46	83.85	74.66	65.74	56.97	48.24	39.41	30.20	19.90	00.00	00.00	00.00
	50	145.39	130.34	118.32	107.78	98.15	89.12	80.51	72.22	64.15	56.24	48.43	40.64	32.80	24.72	15.92	00.00	00.00
	45	132.70	121.23	111.18	101.99	93.38	85.18	77.30	69.66	62.21	54.89	47.69	40.56	33.48	26.40	19.24	11.77	00.01
	40	123.75	114.09	105.25	96.96	89.08	81.49	74.14	66.97	59.94	53.04	46.23	39.49	32.82	26.20	19.62	13.07	6.53
	35	116.66	108.11	100.08	92.44	85.07	77.92	70.94	64.10	57.36	50.71	44.12	37.57	31.05	24.50	17.87	10.93	00.01
	30	110.68	102.87	95.41	88.22	81.22	74.38	67.66	61.02	54.45	47.91	41.38	34.82	28.16	21.25	13.70	00.00	00.00
	25	105.41	98.12	91.06	84.18	77.44	70.80	64.22	57.69	51.16	44.59	37.95	31.12	23.92	15.80	00.00	00.00	00.00
	20	100.62	93.68	86.90	80.23	73.64	67.10	60.57	54.02	47.40	40.64	33.64	26.15	17.52	00.00	00.00	00.00	00.00
	15	96.15	89.44	82.82	76.27	69.74	63.20	56.61	49.92	43.05	35.86	28.09	18.99	00.01	00.00	00.00	00.00	00.00
	10	91.86	85.28	78.75	72.22	65.65	59.01	52.24	45.24	37.88	29.83	20.30	00.10	00.00	00.00	00.00	00.00	00.00
	5	87.66	81.13	74.59	67.99	61.29	54.42	47.30	39.75	31.44	21.50	00.00	00.00	00.00	00.00	00.00	00.00	00.00
	0	83.47	76.90	70.25	63.48	56.52	49.25	41.52	32.95	22.62	00.01	00.00	00.00	00.00	00.00	00.00	00.00	00.00
	-5	79.19	72.48	65.63	58.63	51.16	43.23	34.41	23.69	00.00	00.00	00.00	00.00	00.00	00.00	00.00	00.00	00.00
	-10	74.72	67.78	60.59	53.04	44.91	35.83	24.73	00.00	00.00	00.00	00.00	00.00	00.00	00.00	00.00	00.00	00.00
	-15	69.96	62.64	54.93	46.60	37.25	25.77	00.01	00.00	00.00	00.00	00.00	00.00	00.00	00.00	00.00	00.00	00.00
	-20	64.76	56.86	48.32	38.69	26.81	00.01	00.00	00.00	00.00	00.00	00.00	00.00	00.00	00.00	00.00	00.00	00.00
	-25	58.88	50.10	40.18	27.88	00.00	00.00	00.00	00.00	00.00	00.00	00.00	00.00	00.00	00.00	00.00	00.00	00.00
	-30	51.99	41.74	29.00	00.00	00.00	00.00	00.00	00.00	00.00	00.00	00.00	00.00	00.00	00.00	00.00	00.00	00.00
	-35	43.41	30.20	00.00	00.00	00.00	00.00	00.01	00.00	00.00	00.00	00.00	00.00	00.00	00.00	00.00	00.00	00.00
	-40	31.50	00.00	00.00	00.00	00.00	00.00	00.00	00.00	00.00	00.00	00.00	00.00	00.00	00.00	00.00	00.00	00.00

Relation used: sin (alt) = sin δ sin φ + cos δ cos φ cos HA.

Table 6.4.2. Table of co-ordinates of points for plotting azimuth curves relating to latitutde 40°N for use on Star Charts of various types, Polar, Cartesian, Mercator or Stereographic.

Azimuth

HA \ Az	10	20	30	40	50	60	70	80	90	100	110	120	130	140	150	160	170	180
10	64.66	55.39	50.64	47.64	45.47	43.75	42.27	40.90	39.57	38.18	36.64	34.82	32.46	29.08	23.45	11.51	−24.67	−90.00
20	73.24	63.61	57.37	52.87	49.31	46.30	43.56	40.93	38.26	35.37	32.05	27.96	22.48	14.38	00.87	−23.66	−60.16	−90.00
30	77.28	68.36	61.70	56.39	51.88	47.82	43.96	40.09	36.01	31.45	26.06	19.28	10.15	−2.93	−21.99	−46.85	−71.42	−90.00
40	79.51	71.26	64.50	58.67	53.41	48.42	43.48	38.33	32.73	26.33	18.64	9.00	−3.51	−19.66	−39.03	−58.97	−76.34	−90.00
50	80.85	73.08	66.24	59.99	54.04	48.16	42.09	35.59	28.34	19.95	9.95	−2.18	−16.69	−33.12	−50.02	−65.64	−78.97	−90.00
60	81.67	74.16	67.19	60.49	53.84	47.00	39.73	31.75	22.76	12.42	0.46	−13.12	−27.88	−42.85	−56.98	−69.58	−80.53	−90.00
70	82.14	74.71	67.48	60.24	52.78	44.86	36.26	26.72	16.01	4.04	−9.06	−22.84	−36.59	−49.60	−61.45	−72.03	−81.47	−90.00
80	82.34	74.79	67.14	59.20	50.76	41.60	31.53	20.42	8.29	4.63	−17.86	−30.82	−43.02	−54.20	−64.33	−73.55	−82.03	−90.00
90	82.31	74.42	66.14	57.27	47.61	37.00	25.41	12.96	0.00	−12.96	−25.41	−37.00	−47.61	−57.27	−66.14	−74.42	−82.31	−90.00
100	82.03	73.55	64.33	54.20	43.02	30.82	17.86	4.63	−8.29	−20.42	−31.53	−41.60	−50.76	−59.20	−67.14	−74.79	−82.34	−90.00
110	81.47	72.03	61.45	49.60	36.59	22.84	9.06	−4.04	−16.01	−26.72	−36.26	−44.86	−52.78	−60.24	−67.48	−74.71	−82.14	−90.00
120	80.53	69.58	56.98	42.85	27.88	13.12	−0.46	−12.42	−22.76	−31.75	−39.73	−47.00	−53.84	−60.49	−67.19	−74.16	−81.67	−90.00
130	78.97	65.64	50.02	33.12	16.69	2.18	−9.95	−19.95	−28.34	−35.59	−42.09	−48.16	−54.04	−59.99	−66.24	−73.08	−80.85	−90.00
140	76.34	58.97	39.03	19.66	3.51	−9.00	−18.64	−26.33	−32.73	−38.33	−43.48	−48.42	−53.41	−58.67	−64.50	−71.26	−79.51	−90.00
150	71.42	46.85	21.99	2.93	−10.15	−19.28	−26.06	−31.45	−36.01	−40.09	−43.96	−47.82	−51.88	−56.39	−61.70	−68.36	−77.28	−90.00
160	60.16	23.66	−0.87	−14.38	−22.48	−27.96	−32.05	−35.37	−38.26	−40.93	−43.56	−46.30	−49.41	−52.87	−57.37	−63.61	−73.24	−90.00
170	24.67	−11.51	−23.45	−29.08	−32.46	−34.82	−36.64	−38.18	−39.57	−40.90	−42.27	−43.75	−45.47	−47.64	−50.64	−55.39	−64.66	−90.00
180	−40.00	−40.00	−40.00	−40.00	−40.00	−40.00	−40.00	−40.00	−40.00	−40.00	−40.00	−40.00	−40.00	−40.00	−40.00	−40.00	−40.00	−65.00

Hour Angle (HA)

Table 6.4.3. Table of co-ordinates of points for plotting altitude curves, or Almucantars for latitude 57°N.

Declination (δ)	Altitude																
	5	10	15	20	25	30	35	40	45	50	55	60	65	70	75	80	85
80	0.00	0.00	0.00	0.00	0.00	0.00	0.00	0.00	0.00	129.29	94.11	64.92	31.80	0.00	0.00	0.00	0.00
75	0.00	0.00	0.00	0.00	0.00	0.00	0.00	0.00	136.94	108.21	86.32	66.62	46.96	23.16	0.00	0.00	0.00
70	0.00	0.00	0.00	0.00	0.00	0.00	0.00	141.26	115.77	96.80	80.40	65.27	50.61	35.53	17.32	0.00	0.00
65	0.00	0.00	0.00	0.00	0.00	0.00	144.13	120.64	103.31	88.52	75.13	62.60	50.56	38.71	26.59	12.50	0.00
60	0.00	0.00	0.00	0.00	0.00	146.21	124.12	107.86	94.04	81.61	70.07	59.13	48.63	38.41	28.37	18.33	7.67
55	0.00	0.00	0.00	0.00	147.81	126.77	111.29	98.14	86.31	75.34	64.97	55.03	45.41	36.01	26.76	17.58	8.00
50	0.00	0.00	0.00	149.11	128.90	114.01	101.35	89.95	79.36	69.33	59.69	50.31	41.09	31.89	22.49	12.07	0.00
45	0.00	0.00	150.21	130.68	116.26	103.98	92.90	82.58	72.77	63.30	54.04	44.86	35.56	25.82	14.47	0.00	0.00
40	0.00	151.15	132.20	118.19	106.21	95.38	85.26	75.61	66.25	57.05	47.83	38.41	28.34	16.22	0.00	0.00	0.00
35	151.99	133.55	119.87	108.16	97.52	87.56	78.03	68.74	59.56	50.30	40.72	30.35	17.60	0.00	0.00	0.00	0.00
30	134.77	121.39	109.90	99.43	89.60	80.15	70.91	61.72	52.40	42.69	32.04	18.73	0.00	0.00	0.00	0.00	0.00
25	122.78	111.49	101.17	91.44	82.06	72.85	63.64	54.26	44.40	33.50	19.70	0.00	0.00	0.00	0.00	0.00	0.00
20	112.97	102.78	93.14	83.81	74.62	65.39	55.93	45.93	34.80	20.56	0.00	0.00	0.00	0.00	0.00	0.00	0.00
15	104.30	94.73	85.45	76.26	67.00	57.46	47.34	35.98	21.30	0.00	0.00	0.00	0.00	0.00	0.00	0.00	0.00
10	96.26	87.01	77.82	68.52	58.91	48.65	37.07	22.00	0.00	0.00	0.00	0.00	0.00	0.00	0.00	0.00	0.00
5	88.51	79.32	69.98	60.29	49.89	38.11	22.72	0.00	0.00	0.00	0.00	0.00	0.00	0.00	0.00	0.00	0.00
0	80.79	71.41	61.63	51.10	39.11	23.36	0.00	0.00	0.00	0.00	0.00	0.00	0.00	0.00	0.00	0.00	0.00
−5	72.82	62.95	52.28	41.06	23.99	0.00	0.00	0.00	0.00	0.00	0.00	0.00	0.00	0.00	0.00	0.00	0.00
−10	64.28	53.47	41.06	24.61	0.00	0.00	0.00	0.00	0.00	0.00	0.00	0.00	0.00	0.00	0.00	0.00	0.00
−15	54.67	42.04	25.23	0.00	0.00	0.00	0.00	0.00	0.00	0.00	0.00	0.00	0.00	0.00	0.00	0.00	0.00
−20	43.05	25.87	0.00	0.00	0.00	0.00	0.00	0.00	0.00	0.00	0.00	0.00	0.00	0.00	0.00	0.00	0.00
−25	26.54	0.00	0.00	0.00	0.00	0.00	0.00	0.00	0.00	0.00	0.00	0.00	0.00	0.00	0.00	0.00	0.00
−30	0.00	0.00	0.00	0.00	0.00	0.00	0.00	0.00	0.00	0.00	0.00	0.00	0.00	0.00	0.00	0.00	0.00
−35	0.00	0.00	0.00	0.00	0.00	0.00	0.00	0.00	0.00	0.00	0.00	0.00	0.00	0.00	0.00	0.00	0.00
−40	0.00	0.00	0.00	0.00	0.00	0.00	0.00	0.00	0.00	0.00	0.00	0.00	0.00	0.00	0.00	0.00	0.00

Relation used: $\sin(\text{alt}) = \sin \delta \sin \phi + \cos \delta \cos \phi \cos \text{HA}$.

Table 6.4.4. Table of co-ordinates of points for plotting azimuth curves relating to latitutde 57°N for use on Star Charts of various types, Polar, Cartesian, Mercator or Stereographic.

Azimuth

Hour Angle (HA)	10	20	30	40	50	60	70	80	90	100	110	120	130	140	150	160	170	180
10	73.26	67.32	64.20	62.20	60.73	59.54	58.51	57.55	56.60	55.60	54.47	53.11	51.32	48.66	43.96	32.64	-16.26	-90.00
20	78.71	72.50	68.47	65.51	63.13	61.07	59.17	57.30	55.35	53.19	50.62	47.32	42.62	34.94	19.76	-15.55	-64.69	-90.00
30	81.31	75.46	71.12	67.61	64.58	61.78	59.05	56.23	53.13	49.52	44.98	38.78	29.39	13.47	-14.39	-49.93	-75.52	-90.00
40	82.76	77.26	72.77	68.86	65.26	61.75	58.14	54.22	49.71	44.17	36.87	26.48	10.72	-12.78	-40.85	-64.14	-79.72	-90.00
50	83.64	78.36	73.73	69.44	65.26	60.97	56.34	51.07	44.71	36.57	25.54	10.08	-10.78	-34.47	-55.34	-70.82	-81.85	-90.00
60	84.17	78.99	74.16	69.43	64.58	59.36	53.44	46.41	37.59	26.08	10.82	-8.42	-29.44	-48.37	-63.25	-74.47	-83.09	-90.00
70	84.46	79.25	74.12	68.84	63.14	56.71	49.11	39.72	27.77	12.54	-5.78	-25.15	-42.65	-56.82	-67.89	-76.65	-83.84	-90.00
80	84.57	79.19	73.61	67.57	60.74	52.67	42.78	30.38	14.97	-2.94	-21.34	-37.83	-51.34	-62.09	-70.76	-77.99	-84.28	-90.00
90	84.51	78.79	72.54	65.44	57.01	46.67	33.75	17.94	0.00	-17.94	-33.75	-46.67	-57.01	-65.44	-72.54	-78.79	-84.51	-90.00
100	84.28	77.99	70.76	62.09	51.34	37.83	21.34	2.94	-14.97	-30.38	-42.78	-52.67	-60.74	-67.57	-73.61	-79.19	-84.57	-90.00
110	83.84	76.65	67.89	56.82	42.65	25.15	5.78	-12.54	-27.77	-39.72	-49.11	-56.71	-63.14	-68.84	-74.16	-79.25	-84.57	-90.00
120	83.09	74.47	63.25	48.37	29.44	8.42	-10.82	-26.08	-37.59	-46.41	-53.44	-59.30	-64.58	-69.43	-74.12	-78.99	-84.17	-90.00
130	81.85	70.82	55.34	34.47	10.78	-10.08	-25.54	-36.57	-44.71	-51.07	-56.34	-60.97	-65.26	-69.44	-73.73	-78.36	-83.64	-90.00
140	79.72	64.14	40.85	12.78	-10.72	-26.48	-36.87	-44.17	-49.71	-54.22	-58.14	-61.75	-65.26	-68.86	-72.77	-77.26	-82.76	-90.00
150	75.52	49.93	14.39	-13.47	-29.39	-38.78	-44.98	-49.52	-53.13	-56.23	-59.05	-61.78	-64.58	-67.61	-71.12	-75.46	-81.31	-90.00
160	64.69	15.55	-19.76	-34.94	-42.62	-47.32	-50.62	-53.19	-55.35	-57.30	-59.17	-61.07	-63.13	-65.51	-68.47	-72.50	-78.71	-90.00
170	16.26	-32.64	-43.96	-48.66	-51.32	-53.11	-54.47	-55.60	-56.60	-57.55	-58.51	-59.54	-60.73	-62.20	-64.20	-67.32	-73.26	-90.00
180	-57.00	-57.00	-57.00	-57.00	-57.00	-57.00	-57.00	-57.00	-57.00	-57.00	-57.00	-57.00	-57.00	-57.00	-57.00	-57.00	-57.00	-57.00

Relation used: $\tan \delta = \dfrac{\sin \delta \cos HA + \sin HA \cot (Az)}{\cos \phi}$

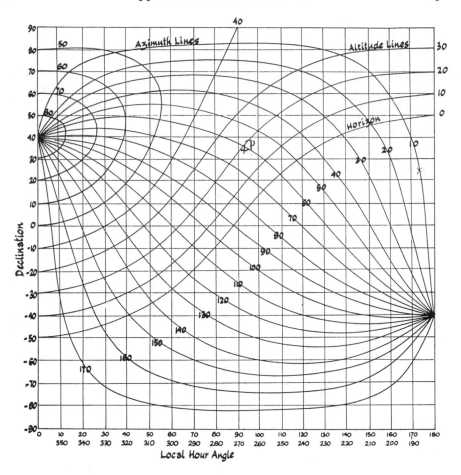

Fig. 6.4.1. Nomogram for Latitude 40°N.

The diagram can be entered with any two of the parameters: Altitude, Azimuth,
Declination, Local Hour Angle to obtain the other two

i.e. LHA with Dec. gives Alt. and Az. or vice versa

Alt. with LHA gives Dec. and Az. or vice versa

LHA with Az. gives Dec. and Alt. or vice versa

Example. For the point P.LHA = 094° or 266° The altitude is 19°

Dec. = 34°   and the azimuth is 61°

To find Local Hour Angle:

Find Local Sidereal time approximately from the equation LST = LMT + (4 minutes
× the number of days since 21st September. Then LHA = LST − Right Ascension.

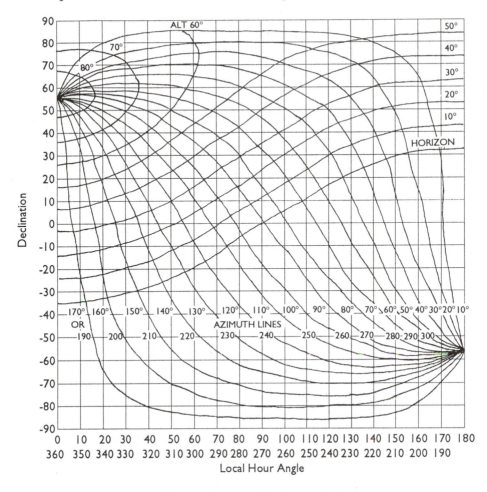

Fig. 6.4.2. Nomogram for Latitude 57° N.

Enter the diagram with LHA and Dec, to find Altitude and Azimuth, or with Altitude and Azimuth to find LHA and Dec.

Example. For the point LHA = 094° or 266° Alt = 26°

Dec. = +34°   Az = 066° or 294°

To Find LHA:

Find Local Sidereal Time approximately from the equation LST = LMT + (number of days since 21st September × 4 minutes). Then LHA = LST − RA.

6.5   THE RIGHT ASCENSIONS AND DECLINATIONS OF SOME OF THE BRIGHTEST STARS

Table 6.5.1

| Star | | RA | | Declination | | Apparent |
Name	Constellation	h	m	°	′	Magnitude
Alpheratz	α Andromeda	0	07.2	+28	58	2.1
Schedar #	α Cassiopeiae	0	39.2	+56	25	2.3
Mirach	α Andromeda	1	08.4	+35	30	2.4
Achernar	α Eridani	1	36.9	−57	21	0.6
Polaris #	α Ursae Minoris	2	09.2	+89	10	2.1
Algol	β Persei	3	06.7	+40	52	Var
Mirfak	α Persei	3	22.7	+49	47	1.9
Aldebaran	α Tauri	4	34.6	+16	28	1.1
Rigel	β Orionis	5	13.4	−8	14	0.3
Capella	α Aurigae	5	15.0	+45	59	0.2
Bellatrix	γ Orionis	5	23.9	+6	20	1.7
Betelgeuse	α Orionis	5	53.9	+7	24	Var
Mirzam	β Canis Majoris	6	21.7	−17	57	2.0
Canopus	α Carinae	6	23.4	−52	41	−0.9
Sirius	α Canis Majoris	6	44.1	−16	41	−1.6
Castor	α Geminorum	7	33.1	+31	56	1.6
Procyon	α Canis Minoris	7	38.1	+5	17	0.5
Pollux	β Geminorum	7	43.9	+28	05	1.2
Regulus	α Leonis	10	07.1	+12	05	1.3
Merak #	β Ursae Majoris	11	00.5	+56	30	2.4
Dubhe #	α Ursae Majoris	11	02.3	+61	53	1.9
Denebola	β Leonis	11	47.9	+14	42	2.2
	α Crucis†	12	25.3	−62	58	1.0
	γ Crucis†	12	29.9	−56	59	1.6
	β Crucis†	12	46.4	−59	34	1.5
Spica	α Virginis	13	24.0	−11	02	1.2
Hadar	β Centauri†	14	02.2	−60	16	0.9
Arctrus	α Bootis	14	14.6	+19	18	0.2
Rigil Kent	α Centauri†	14	38.0	−60	44	0.1
Alphecca	α Coronae Borealis	15	33.7	+26	47	2.3
Antares	α Scorpii	16	28.0	−26	23	1.2
	α Trianguli Australis†	16	46.2	−68	59	1.9
Rasalhague	α Ophiuchi	17	33.9	+12	35	2.1
Vega	α Lyrae	18	36.2	+38	46	0.1
Altair	α Aquillae	19	49.7	+8	48	0.9
Deneb	α Cygni	20	40.6	+45	12	1.3
Fomalhaut	α Piscis Austrini	22	56.4	−29	45	1.3

In northern latitudes where ϕ is positive, all stars having a declination between +90° and ϕ −90° can appear above the horizon. For example in latitude 51°N all stars having declinations between 90°N and −39°S can so appear. The six stars marked † in the table, Crucis α, β and γ, Centauri α and β, and Trianguli Australis α cannot appear to observers in latitude 51°N.

In southern latitudes where ϕ is negative, all stars having a declination between −90°S and 90° + ϕ are visible, i.e. can appear above the horizon. For example in latitude −35° (southern Australia) only those stars with declinations between −90°S and 55°N can be observed. Only four of the listed bright stars are thus excluded from an observer in this southern latitude, and are marked # (Schedar, Polaris, Merak and Dubhe).

6.5.2 Stellar Distances

1 parsec is a distance of 3.2616 light years

10 parsecs is a distance of 32.616 light years

A parsec is not an angle but is that distance, D, of a star which gives rise to a parallax of 1".

$$\text{Parallax in radians} = \frac{\text{distance Earth to Sun (r)}}{\text{distance of star (D)}}$$

This expressed in seconds of arc =

$$\frac{r}{D} \times 57.295\ 78 \times 60 \times 60''$$

If parallax angle is 1" then,

$$1 = \frac{r}{D} \times 2.062\ 65 \times 10^5$$

but,

$$r = 1.496\ 00 \times 10^{11}\ \text{m}$$

and

$$D = 1.4960 \times 10^{11} \times 2.062\ 65 \times 10^5\ \text{m}$$

$$1\ \text{light year} = 9.4607 \times 10^{15}\ \text{m}$$

$$D, \text{ or } 1\ \text{parsec} = \frac{1.496 \times 10^{11} \times 2.062\ 65 \times 10^5}{9.4607 \times 10^{15}}$$

$$= 3.2616\ \text{light years}$$

The distance of a star in parsecs is the reciprocal of its parallax in seconds of arc. If, for example, the parallax of a star is 0.026", its distance in parsecs is the reciprocal of 0.026,

$$\frac{1}{0.026} = 38.461\ \text{parsecs}$$

6.6   USEFUL INFORMATION

1. **The Sun**
    Solar mass 1.990×10^{30} kg
    Solar radius 6.960×10^{8} m
    Semi-diameter at
      mean distance $15'\ 59''$
    Astronomical unit (1AU) 1.496×10^{11} m

2. **The Earth**
    Earth mass 5.976×10^{24} kg
    Polar radius 6.356775×10^{6} m
    Equatorial radius 6.378160×10^{6} m

3. **The Moon**
    Lunar mass 7.35×10^{22} kg
    Lunar radius 1.738×10^{6} m
    Semi-diameter at
      mean distance $15'\ 32''.6$ (geocentric)
    Mean equatorial
      horizontal parallax $3422''.54$
    Mean distance
    from Earth 3.8440×10^{8} m
    Inclination of orbit
    to ecliptic $5°\ 8'\ 43''$

4. **Time**
    24^{h} mean solar time $24^{h}\ 03^{m}\ 56^{s}$
                                                      mean sidereal time

    24^{h} mean sidereal time $23^{h}\ 56^{m}\ 04^{s}.09$
      mean solar time
    Mean solar day $24^{h}\ 03^{m}\ 56^{s}.555$
                                                    $1^{d}.00273791$
                                                      mean sidereal time

    Mean sidereal day $23^{h}\ 56^{m}\ 4^{s}.091$
      mean solar time

5. **Length of Year**
    Julian   365.25 days
    Tropical   365.24219 days
       (equinox to equinox)
    Sidereal   365.25636 days
       (fixed star to fixed star)
    Anomalistic   365.25964 days
       (perihelion to perihelion)
    Eclipse   346.62003 days
       (lunar node to lunar node)

6. **Length of Month**
    Synodic   29.53059 days
       (new moon to new moon)
    Tropical   27.32158 days
       (equinox to equinox)
    Sidereal   27.32166 days
       (fixed star to fixed star)
    Anomalistic   27.55455 days
       (perigee to perigee)

7. **Miscellaneous**
    Gravitational constant $6.670 \times 10^{-11} \, kg^{-1} \, m^3 \, s^{-2}$
    Speed of light $2.997925 \times 10^8 \, m \, s^{-1}$
    Parsec   206264.8 AU
                                        3.0857×10^{16} m
                                        3.2616 light years
    Light year 9.4607×10^{15} m
                                        6.324×10^4 AU
    Nautical mile   1.853 km
                                        1853 m
    Annual general $50''.2564 + 0'',0222 \, T$
       precession   (T measured in Julian
                                        centuries from 1900)
    1 radian $57°.29578$
    1°   0.01745 radians

## 6.7	SUGGESTED FURTHER READING AND SOURCES OF INFORMATION

Suggested Further Reading

The Handbook of the British Astronomical Association and *An Explanatory Suplement.* Published Annually by The British Astronomical Association,

Whitaker's Almanac. The Shorter Edition, Published Annually by J. Whitaker & Sons Ltd., 12 Dyott Street, London WC1A 1DF.

Nuffield Physics Teacher's Guide _ Astronomy in the Science Syllabus. Longmans,

Hogben, L. (1956) *Science for the Citizen.* George Allen & Unwin, London.

Brewer, S. G. (1988) *D.I.Y. Astronomy.* Edinburgh University Press, Edinburgh.

Moore, P. (Annual) *Year Book of Astronomy.* Sedgwick & Jackson, London.

Moore, P. (1979) *The Guinness Book of Astronomy Facts and Feats.* Guinness Superlatives Ltd. Enfield Middlesex.

Hoyle, Fred. (1977) *On Stonehenge.* Heinemann Educational Books Ltd.

Ronan, C. A. (1981) *The Practical Astronomer.* Pan Books, London.

Ridpath, I. (1989) *Norton's 2000 Star Atlas and Reference Handbook.* Longman Scientific & Technical, London.

Couper, H. (1980) *Exploring Space.* Marks & Spencer, London.

Brown, H. (1978) *Man and the Stars.* Oxford University Press, Oxford.

Roy, A. E. & Clarke, D. (1977) *Astronomy Principles and Practice.* Adam Hilger Ltd., Bristol.

McAleer, N. (1982) *The Cosmic Mind Bogling Book.* Hodder & Stoughton, London.

Handbook for Astronomical Societies. Federation of Astronomical Societies.

Journal of the British Astronomical Association. First published in 1890, now published bi-monthly.

Astronomy Now. Monthly, Intra Press, London.

Gnomen, The Newsletter of the Association for Astronomy Education.

Popular Astronomy. Journal of the Junior Astronomical Society.

Ridpath, I. (19) *The Young Astronomer's Handbook.*

Sky and Telescope. Sky Publishing Corporation U.S.A. (monthly).

Gnomon. Newsletter of the Association for Astronomy Eduction.

The Bulletin. The British Sundial Society.

6.8   TAILPIECE

Finally, as at the beginning, (1.1) ask questions.

Many skywatchers from their early days soon become filled with ''Satiable Curiosity'' (Kipling's Elephant Child.), but for fear of displaying ignorance among their experienced fellow members of their class or astronomical Society, hesitate to put the questions they long to ask, so that they remain unanswered for years. So far, you may have acquired answers only to a small fraction of the questions that regular sky watchers ask. Take heart, even our most able professional astronomers are similarly situated; but the satisfying thing to do is to keep asking questions and seeking the answers, thus emulating the young lady in Kipling's concluding lines of the little rhyme we began with;

''I know a person small -
She keeps ten million serving men,
Who get not rest at all ! !

She sends them abroad on her own affairs,
From the second she opens her eyes -
One million Hows, two million Wheres,
And seven million Whys !''

Index

Altitude-Azimuth overlays, 121
Altazimeter, 33, 35
Altitude of the Pole Star, 16
Astrolabe, 130
Axis of the Earth, 17, 194
Azimuth sundial, 86
Advice for beginners, 215

Barlow lens, 164
Barycentre, 193
Binoculars, observing with ease and comfort, 167, 163
Blue of the sky, 100
Brightest stars, 228

Calendars and the moon, 63
Cats eyes, 13
Celestial sphere, and terms used, 19, 53, 57
Centripetal force, 63
Chaucer, CANTERBURY Tales, Pillar Sundial, 82
Compass bearings, Azimuths and the false watch compass, 101

Dispersion of light through a prism, 143
Distances of stars, 229

Earth's axis, 17, 194
Earth globe, 16
Earth's radius, and its measurement, 55, 201
Eccentricity of the Earth's orbit, 96
Ecliptic and its plane, 60, 194
Equation of Time, 94
Eyepiece magnification, 16
Eyepiece projection for photography, 158, 160, 162

Field of view, of eyepiece and telescope, 158, 162
Foucault Pendulum, 181
Further reading, 232

Galileo, 66
Geocentric view of Solar system, 65
Geostationary orbit, 185
Glossary of terms used in Astronomy, 218
Graphs and nomograms, 221
Gravity, 64, 178, 179
Graticules, 121

Horizon curve in planispheres, 120
Hour angle, 54

Latitude, 16
Lens formulae, 258
        refraction, 140
Light pollution, 13, 173

Magnitudes of stars, 199, 200, 228
Measurements in Astronomy, 25
Measuring angles, Altitudes and Azimuths, 25, 152
Meteorites, 183
Molecules in the atmosphere, 202
Moon and the Calendar, 63, 110
Moon and its magnification by telescope, 157

Nomograms, 114, 134, 135, 221, 235, 236
Noon marks, 99
Number of stars that can be seen by unaided eye, 14

Parabolic Mirrors, 203
Paths of stars across the sky, 137
Photography with the telescope, 158, 160

Pillar dial Sundial, 78
Planetarium, value of, 39
Planispheres, 114, 121, 125
Playground Sundial an Azimuth Dial, 86
Pocket Star Globe, 23, 106
Pole star, Polaris, 16
Postscript, 233
Precession of Earth's axis, 194
Printouts for planisphere graticules, 118, 119

Quadrant Altazimeter, 29

Radiations, 139
Radius of the Earth, 55, 201
Rainbows, 170
Refractions, 140
Resolution of a telescope, 164, 176, 177
Risings and settings of celestial bodies, 102

Satellites in orbit, 185
Setting circles, 154, 155
Sextant model, 31
Sidereal Time and star clock, 39, 47, 49
Solar system heliocentric view, 52, 63
Southern Hemisphere Stars, 198, 228
Spectra, 174
Spherical Trigonometry, 114
Spherical Trigonometrical Formulae, 195, 198
Star Globes, 18, 19, 21, 23, 41
Stars of the month and their special projection, 131

Star Trails, 42, 137
Stonehenge — a Solar Lunar Observatory, 110
Sundials, 63, 71, 83, 86, 94, 104, 106
Sun's special nomogram, 136
Sun's path in the sky and how it changes during the year, 137
Sun Spots, 66
Sunrise and Sunset, 102

Tables of coordinates for Alt-Az curves for Latitudes 40° and 57°, 222–225
Telescopes, 144, 154, 176
    Lightgather power, 145, 176
    Magnification, 176
    Mountings, 148, 150
    Refractors and reflectors, 167
    Using telescopes, a local set square, 154
    What your telescope can do under good conditions, 154, 176
    Resolution, 164, 176, 177
Time by the Stars, 44, 46, 63
Tips on giving a talk, 215
Transits, and Sidereal Time, 31, 36, 98

Useful information, 230

Venus, measuring its elongation, 50

Watch Compass and its error, 101